# Handbook of Pollution: Air, Noise and Water

# Handbook of Pollution: Air, Noise and Water

Edited by Gregory Rawlings

Syrawood Publishing House
New York

Published by Syrawood Publishing House,
750 Third Avenue, 9th Floor,
New York, NY 10017, USA
www.syrawoodpublishinghouse.com

**Handbook of Pollution: Air, Noise and Water**
Edited by Gregory Rawlings



International Standard Book Number: 978-1-68286-617-7 (Hardback)



**Cataloging-in-Publication Data**

Handbook of pollution : air, noise and water / edited by Gregory Rawlings.
p. cm.
Includes bibliographical references and index.
ISBN 978-1-68286-617-7
1. Pollution. 2. Air--Pollution. 3. Noise pollution. 4. Water--Pollution.
I. Rawlings, Gregory.
TD174 .H36 2018
628.5--dc23

Printed in China.

# TABLE OF CONTENTS

# PREFACE

The rise in population and industries has led to many problems and most critical out of them is pollution. Due to the increase in the number of vehicles, growing infrastructure, rapid deforestation, emission of poisonous gases and excessive use of machines and technological devices like ACs, has resulted in an increase in the pollution of air, noise and water. It has led to many deadly skin and lung diseases. This book is designed to present the subject of pollution in an in-depth manner. This textbook aims to serve as a resource guide for students and experts alike and contribute to the growth of the discipline.

A foreword of all chapters of the book is provided below:

**Chapter 1 -** Pollutants are the substances in our environment that cause harmful effects. The effect they have can either be long term or short term in nature. The various types of pollutants of natural resources are stock pollutants, notable pollutants, light pollution and fund pollutants. This chapter will provide an integrated understanding of pollutant and pollution; **Chapter 2 -** Pollution can damage the environment immensely. There are various types of pollutions such as noise pollution, water pollution, air pollution, light pollution and toxic hotspot. The increase in pollution has resulted in the destruction of habitats and extinction of several species. Air, noise and water pollution can best understood in confluence with the major topics listed in the following chapter; **Chapter 3 -** It is very important to control pollution to sustain the life of our planet. Some of the devices and technologies introduced for air and water pollution control are dust collector, scrubber, sewage treatment and vapor recovery. The chapter serves as a source to understand the devices and technologies used for controlling pollution; **Chapter 4 -** One of the practices used for preventing pollution is recycling. It is the reduction of waste by reusing products and materials and converting waste into useful products. The other methods of reducing pollution are waste minimization, pollution prevention, regulation and monitoring of pollution, etc. This chapter will help the readers in developing a better understanding of air, noise and water pollution.

At the end, I would like to thank all the people associated with this book devoting their precious time and providing their valuable contributions to this book. I would also like to express my gratitude to my fellow colleagues who encouraged me throughout the process.

**Editor**

1

# An Introduction to Pollutant and Pollution

Pollutants are the substances in our environment that cause harmful effects. The effect they have can either be long term or short term in nature. The various types of pollutants of natural resources are stock pollutants, notable pollutants, light pollution and fund pollutants. This chapter will provide an integrated understanding of pollutant and pollution.

## Pollutant

Surface runoff, also called nonpoint source pollution, from a farm field in Iowa, United States during a rain storm. Topsoil as well as farm fertilizers and other potential pollutants run off unprotected farm fields when heavy rains occur.

A pollutant is a substance or energy introduced into the environment that has undesired effects, or adversely affects the usefulness of a resource. A pollutant may cause long- or short-term damage by changing the growth rate of plant or animal species, or by interfering with human amenities, comfort, health, or property values. Some pollutants are biodegradable and therefore will not persist in the environment in the long term. However, the degradation products of some pollutants are themselves polluting such as the products DDE and DDD produced from degradation of DDT.

### Different Types of Pollutants in Nature

### Stock Pollutants

Pollutants, towards which the environment has little or no absorptive capacity are called *stock pollutants*. (e.g. persistent synthetic chemicals, non-biodegradable plastics, and heavy metals). Stock pollutants accumulate in the environment over time. The damage they cause increases as more pollutant is emitted, and persists as the pollutant accumulates. Stock pollutants can create a burden for future generations, by passing on damage that persists well after the 'benefits' received from incurring that damage; have been forgotten.

## Fund Pollutants

Fund pollutants are those for which the environment has some absorptive capacity. Fund pollutants do not cause damage to the environment unless the emission rate exceeds the receiving environment's absorptive capacity (e.g. carbon dioxide, which is absorbed by plants and oceans). Fund pollutants are not destroyed, but rather converted into less harmful substances, or diluted/ dispersed to non-harmful concentrations.

## Notable Pollutants

Notable pollutants include the following groups:

- Persistent organic pollutants POP
- Environmental Persistent Pharmaceutical Pollutants EPPP
- Polycyclic aromatic hydrocarbons
- Volatile organic compounds

## Light Pollution

Light pollution is the impact that anthropogenic light has on the visibility of the night sky. It also encompasses ecological light pollution which describes the effect of artificial light on individual organisms and on the structure of ecosystems as a whole.

The night sky viewed from Luhasoo bog, Estonia with light pollution in background.

## Zones of Influence

Pollutants can also be defined by their zones of influence, both horizontally and vertically.

## Horizontal Zone

The horizontal zone refers to the area that is damaged by a pollutant. Local pollutants cause damage near the emission source. Regional pollutants cause damage further from the emission source.

## Vertical Zone

The vertical zone is referred to whether the damage is ground-level or atmospheric. Surface pollutants cause damage by concentrations of the pollutant accumulating near the Earth's surface Global pollutants cause damage by concentrations in the atmosphere.

## Regulation

### International

Pollutants can cross international borders and therefore international regulations are needed for their control. The Stockholm Convention on Persistent Organic Pollutants, which entered into force in 2004, is an international legally binding agreement for the control of persistent organic pollutants. Pollutant Release and Transfer Registers (PRTR) are systems to collect and disseminate information on environmental releases and transfers of toxic chemicals from industrial and other facilities.

### European Union

The European Pollutant Emission Register is a type of PRTR providing access to information on the annual emissions of industrial facilities in the Member States of the European Union, as well as Norway.

### United States

Clean Air Act standards. Under the Clean Air Act, the National Ambient Air Quality Standards (NAAQS) are developed by the Environmental Protection Agency (EPA) for six common air pollutants, also called "criteria pollutants": particulates; smog and ground-level ozone; carbon monoxide; sulfur oxides; nitrogen oxides; and lead. The National Emissions Standards for Hazardous Air Pollutants are additional emission standards that are set by EPA for toxic air pollutants.

Clean Water Act standards. Under the Clean Water Act, EPA promulgated national standards for municipal sewage treatment plants, also called *publicly owned treatment works,* in the *Secondary Treatment Regulation.* National standards for industrial dischargers are called *Effluent guidelines* (for existing sources) and New Source Performance Standards, and currently cover over 50 industrial categories. In addition, the Act requires states to publish water quality standards for individual water bodies to provide additional protection where the national standards are insufficient.

RCRA standards. The Resource Conservation and Recovery Act (RCRA) regulates the management, transport and disposal of municipal solid waste, hazardous waste and underground storage tanks.

A tank or piping network that has at least 10 percent of its volume underground is known as an underground storage tank (UST). They often store substances such as petroleum, that are harmful to the surrounding environment should it become contaminated.

## Persistent Organic Pollutant

Persistent organic pollutants (POPs) are organic compounds that are resistant to environmental degradation through chemical, biological, and photolytic processes. Because of their persistence, POPs bioaccumulate with potential significant impacts on human health and the environment.

The effect of POPs on human and environmental health was discussed, with intention to eliminate or severely restrict their production, by the international community at the Stockholm Convention on Persistent Organic Pollutants in 2001.

Many POPs are currently or were in the past used as pesticides, solvents, pharmaceuticals, and industrial chemicals. Although some POPs arise naturally, for example volcanoes and various biosynthetic pathways, most are man-made via total synthesis.

## Consequences of Persistence

POPs typically are halogenated organic compounds and as such exhibit high lipid solubility. For this reason, they bioaccumulate in fatty tissues. Halogenated compounds also exhibit great stability reflecting the nonreactivity of C-Cl bonds toward hydrolysis and photolytic degradation. The stability and lipophilicity of organic compounds often correlates with their halogen content, thus polyhalogenated organic compounds are of particular concern. They exert their negative effects on the environment through two processes, long range transport, which allows them to travel far from their source, and bioaccumulation, which reconcentrates these chemical compounds to potentially dangerous levels. Compounds that make up POPs are also classed as PBTs (Persistent, Bioaccumulative and Toxic) or TOMPs (Toxic Organic Micro Pollutants).

## Long-range Transport

POPs enter the gas phase under certain environmental temperatures and volatize from soils, vegetation, and bodies of water into the atmosphere, resisting breakdown reactions in the air, to travel long distances before being re-deposited. This results in accumulation of POPs in areas far from where they were used or emitted, specifically environments where POPs have never been introduced such as Antarctica, and the Arctic circle. POPs can be present as vapors in the atmosphere or bound to the surface of solid particles. POPs have low solubility in water but are easily captured by solid particles, and are soluble in organic fluids (oils, fats, and liquid fuels). POPs are not easily degraded in the environment due to their stability and low decomposition rates. Due to this capacity for long-range transport, POP environmental contamination is extensive, even in areas where POPs have never been used, and will remain in these environments years after restrictions implemented due to their resistance to degradation.

## Bioaccumulation

Bioaccumulation of POPs is typically associated with the compounds high lipid solubility and ability to accumulate in the fatty tissues of living organisms for long periods of time. Persistent chemicals tend to have higher concentrations and are eliminated more slowly. Dietary accumulation or bioaccumulation is another hallmark characteristic of POPs, as POPs move up the food chain, they increase in concentration as they are processed and metabolized in certain tissues of organisms. The natural capacity for animals gastrointestinal tract concentrate ingested chemicals, along with poorly metabolized and hydrophobic nature of POPs makes such compounds highly susceptible to bioaccumulation. Thus POPs not only persist in the environment, but also as they are taken in by animals they bioaccumulate, increasing their concentration and toxicity in the environment.

## Stockholm Convention on Persistent Organic Pollutants

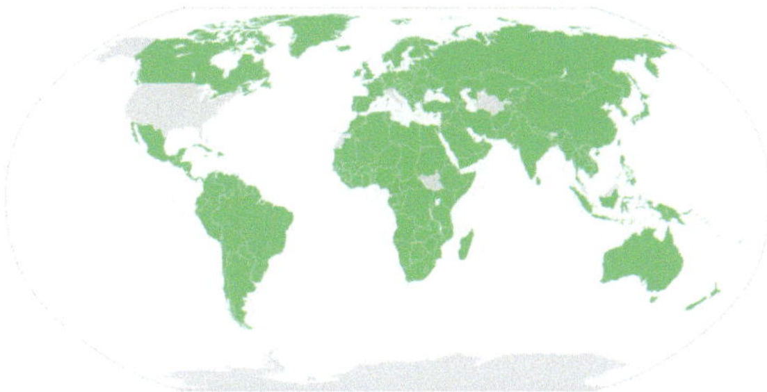

State parties to the Stockholm Convention on Persistent Organic Pollutants

The Stockholm Convention was adopted and put into practice by the United Nations Environment Programme (UNEP) on May 22, 2001. The UNEP decided that POP regulation needed to be addressed globally for the future. The purpose statement of the agreement is "to protect human health and the environment from persistent organic pollutants." As of 2014, there are 179 countries in compliance with the Stockholm convention. The convention and its participants have recognized the potential human and environmental toxicity of POPs. They recognize that POPs have the potential for long range transport and bioaccumulation and biomagnification. The convention seeks to study and then judge whether or not a number of chemicals that have been developed with advances in technology and science can be categorized as POPs or not. The initial meeting in 2001 made a preliminary list, termed the "dirty dozen," of chemicals that are classified as POPs. As of 2014, the United States of America has signed the Stockholm Convention but has not ratified it. There are a handful of other countries that have not ratified the convention but most countries in the world have ratified the convention.

## Compounds on the Stockholm Convention list

In May 1995, the United Nations Environment Programme Governing Council investigated POPs. Initially the Convention recognized only twelve POPs for their adverse effects on human health and the environment, placing a global ban on these particularly harmful and toxic compounds and requiring its parties to take measures to eliminate or reduce the release of POPs in the environment.

1. Aldrin, an insecticide used in soils to kill termites, grasshoppers, Western corn rootworm, and others, is also known to kill birds, fish, and humans. Humans are primarily exposed to aldrin through dairy products and animal meats.

2. Chlordane, an insecticide used to control termites and on a range of agricultural crops, is known to be lethal in various species of birds, including mallard ducks, bobwhite quail, and pink shrimp; it is a chemical that remains in the soil with a reported half-life of one year. Chlordane has been postulated to affect the human immune system and is classified as a possible human carcinogen. Chlordane air pollution is believed the primary route of humane exposure.

3. Dieldrin, a pesticide used to control termites, textile pests, insect-borne diseases and insects living in agricultural soils. In soil and insects, aldrin can be oxidized, resulting in rapid conversion to dieldrin. Dieldrin's half-life is approximately five years. Dieldrin is highly toxic to fish and other aquatic animals, particularly frogs, whose embryos can develop spi-

nal deformities after exposure to low levels. Dieldrin has been linked to Parkinson's disease, breast cancer, and classified as immunotoxic, neurotoxic, with endocrine disrupting capacity. Dieldrin residues have been found in air, water, soil, fish, birds, and mammals. Human exposure to dieldrin primarily derives from food.

4. Endrin, an insecticide sprayed on the leaves of crops, and used to control rodents. Animals can metabolize endrin, so fatty tissue accumulation is not an issue, however the chemical has a long half-life in soil for up to 12 years. Endrin is highly toxic to aquatic animals and humans as a neurotoxin. Human exposure results primarily through food.

5. Heptachlor, a pesticide primarily used to kill soil insects and termites, along with cotton insects, grasshoppers, other crop pests, and malaria-carrying mosquitoes. Heptachlor, even at every low doses has been associated with the decline of several wild bird populations – Canada geese and American kestrels. In laboratory tests have shown high-dose heptachlor as lethal, with adverse behavioral changes and reduced reproductive success at low-doses, and is classified as a possible human carcinogen. Human exposure primarily results from food.

6. Hexachlorobenzene (HCB), was first introduced in 1945–1959 to treat seeds because it can kill fungi on food crops. HCB-treated seed grain consumption is associated with photosensitive skin lesions, colic, debilitation, and a metabolic disorder called porphyria turcica, which can be lethal. Mothers who pass HCB to their infants through the placenta and breast milk had limited reproductive success including infant death. Human exposure is primarily from food.

7. Mirex, an insecticide used against ants and termites or as a flame retardant in plastics, rubber, and electrical goods. Mirex is one of the most stable and persistent pesticides, with a half-life of up to 10 years. Mirex is toxic to several plant, fish and crustacean species, with suggested carcinogenic capacity in humans. Humans are exposed primarily through animal meat, fish, and wild game.

8. Toxaphene, an insecticide used on cotton, cereal, grain, fruits, nuts, and vegetables, as well as for tick and mite control in livestock. Widespread toxaphene use in the US and chemical persistence, with a half-life of up to 12 years in soil, results in residual toxaphene in the environment. Toxaphene is highly toxic to fish, inducing dramatic weight loss and reduced egg viability. Human exposure primarily results from food. While human toxicity to direct toxaphene exposure is low, the compound is classified as a possible human carcinogen.

9. Polychlorinated biphenyls (PCBs), used as heat exchange fluids, in electrical transformers, and capacitors, and as additives in paint, carbonless copy paper, and plastics. Persistence varies with degree of halogenation, an estimated half-life of 10 years. PCBs are toxic to fish at high doses, and associated with spawning failure at low doses. Human exposure occurs through food, and is associated with reproductive failure and immune suppression. Immediate effects of PCB exposure include pigmentation of nails and mucous membranes and swelling of the eyelids, along with fatigue, nausea, and vomiting. Effects are transgenerational, as the chemical can persist in a mother's body for up to 7 years, resulting in developmental delays and behavioral problems in her children. Food contamination has led to large scale PCB exposure.

10. Dichlorodiphenyltrichloroethane (DDT) is probably the most infamous POP. It was widely used as insecticide during WWII to protect against malaria and typhus. After the war, DDT was used as an agricultural insecticide. In 1962, the American biologist Rachel Carson published Silent Spring, describing the impact of DDT spraying on the US environment and human health. DDT's persistence in the soil for up to 10–15 years after application has resulted in widespread and persistent DDT residues throughout the world including the arctic, even though it has been banned or severely restricted in most of the world. DDT is toxic to many organisms including birds where it is detrimental to reproduction due to eggshell thinning. DDT can be detected in foods from all over the world and food-borne DDT remains the greatest source of human exposure. Short-term acute effects of DDT on humans are limited, however long-term exposure has been associated with chronic health effects including increased risk of cancer and diabetes, reduced reproductive success, and neurological disease.

11. Dioxins are unintentional by-products of high-temperature processes, such as incomplete combustion and pesticide production. Dioxins are typically emitted from the burning of hospital waste, municipal waste, and hazardous waste, along with automobile emissions, peat, coal, and wood. Dioxins have been associated with several adverse effects in humans, including immune and enzyme disorders, chloracne, and are classified as a possible human carcinogen. In laboratory studies of dioxin effects an increase in birth defects and stillbirths, and lethal exposure have been associated with the substances. Food, particularly from animals, is the principal source of human exposure to dioxins.

12. Polychlorinated dibenzofurans are by-products of high-temperature processes, such as incomplete combustion after waste incineration or in automobiles, pesticide production, and polychlorinated biphenyl production. Structurally similar to dioxins, the two compounds share toxic effects. Furans persist in the environment and classified as possible human carcinogens. Human exposure to furans primarily results from food, particularly animal products.

## New POPs on the Stockholm Convention List

Since 2001, this list has been expanded to include some polycyclic aromatic hydrocarbons (PAHs), brominated flame retardants, and other compounds. Additions to the initial 2001 Stockholm Convention list are as following POPs:

- Chlordecone, a synthetic chlorinated organic compound,is primarily used as an agricultural pesticide, related to DDT and Mirex. Chlordecone is toxic to aquatic organisms, and classified as a possible human carcinogen. Many countries have banned chlordecone sale and use, or intend to phase out stockpiles and wastes.

- α-Hexachlorocyclohexane (α-HCH) and β-Hexachlorocyclohexane (β-HCH) are insecticides as well as by-products in the production of lindane. Large stockpiles of HCH isomers exist in the environment. α-HCH and β-HCH are highly persistent in the water of colder regions. α-HCH and β-HCH has been linked Parkinson's and Alzheimer's disease.

- Hexabromodiphenyl ether (hexaBDE) and heptabromodiphenyl ether (heptaBDE) are main components of commercial octabromodiphenyl ether (octaBDE). Commercial octaB-

DE is highly persistent in the environment, whose only degradation pathway is through debromination and the production of bromodiphenyl ethers, which can increase toxicity.

- Lindane (γ-hexachlorocyclohexane), a pesticide used as a broad spectrum insecticide for seed, soil, leaf, tree and wood treatment, and against ectoparasites in animals and humans (head lice and scabies). Lindane rapidly bioconcentrates. It is immunotoxic, neurotoxic, carcinogenic, linked to liver and kidney damage as well as adverse reproductive and developmental effects in laboratory animals and aquatic organisms. Production of lindane unintentionally produces two other POPs α-HCH and β-HCH.
- Pentachlorobenzene (PeCB), is a pesticide and unintentional byproduct. PeCB has also been used in PCB products, dyestuff carriers, as a fungicide, a flame retardant, and a chemical intermediate. PeCB is moderately toxic to humane, while highly toxic to aquatic organisms.
- Tetrabromodiphenyl ether (tetraBDE) and pentabromodiphenyl ether (pentaBDE) are industrial chemicals and the main components of commercial pentabromodiphenyl ether (pentaBDE). PentaBDE has been detected in humans in all regions of the world.
- Perfluorooctanesulfonic acid (PFOS) and its salts are used in the production of fluoropolymers. PFOS and related compounds are extremely persistent, bioaccumulating and biomagnifying. The negative effects of trace levels of PFOS have not been established.
- Endosulfans are insecticides to control pests on crops such coffee, cotton, rice and sorghum and soybeans, tsetse flies, ectoparasites of cattle. They are used as a wood preservative. Global use and manufacturing of endosulfan has been banned under the Stockholm convention in 2011, although many countries had previously banned or introduced phase-outs of the chemical when the ban was announced. Toxic to humans and aquatic and terrestrial organisms, linked to congenital physical disorders, mental retardation, and death. Endosulfans' negative health effects are primarily liked to its endocrine disrupting capacity acting as an antiandrogen.
- Hexabromocyclododecane (HBCD) is a brominated flame retardant primarily used in thermal insulation in the building industry. HBCD is persistent, toxic and ecotoxic, with bioaccumulative and long-range transport properties.

## Additive and Synergistic Effects

Evaluation of the effects of POPs on health is very challenging in the laboratory setting. For example, for organisms exposed to a mixture of POPs, the effects are assumed to be additive. Mixtures of POPs can in principle produce synergistic effects. With synergistic effects, the toxicity of each compound is enhanced (or depressed) by the presence of other compounds in the mixture. When put together, the effects can far exceed the approximated additive effects of the POP compound mixture.

## Health Effects

POP exposure may cause developmental defects, chronic illnesses, and death. Some are carcinogens per IARC, possibly including breast cancer. Many POPs are capable of endocrine disruption within the reproductive system, the central nervous system, or the immune system. People and animals are exposed to POPs mostly through their diet, occupationally, or while growing in the womb. For humans not exposed to POPs through accidental or occupational means, over 90% of exposure comes from animal product foods due to bioaccumulation in fat tissues and bioaccumulate through the food chain. In general, POP serum levels increase with age and tend to be higher in females than males.

Studies have investigated the correlation between low level exposure of POPs and various diseases. In order to assess disease risk due to POPs in a particular location, government agencies may produce a human health risk assessment which takes into account the pollutants' bioavailability and their dose-response relationships.

## Endocrine Disruption

The majority of POPs are known to disrupt normal functioning of the endocrine system, for example all of the dirty dozen are endocrine disruptors. Low level exposure to POPs during critical developmental periods of fetus, newborn and child can have a lasting effect throughout its lifespan. A 2002 study synthesizes data on endocrine disruption and health complications from exposure to POPs during critical developmental stages in an organism's lifespan. The study aimed to answer the question whether or not chronic, low level exposure to POPs can have a health impact on the endocrine system and development of organisms from different species. The study found that exposure of POPs during a critical developmental time frame can produce a permanent changes in the organisms path of development. Exposure of POPs during non-critical developmental time frames may not lead to detectable diseases and health complications later in their life. In wildlife, the critical development time frames are in utero, in ovo, and during reproductive periods. In humans, the critical development timeframe is during fetal development.

## Reproductive System

The same study in 2002 with evidence of a link from POPs to endocrine disruption also linked low dose exposure of POPs to reproductive health effects. The study stated that POP exposure can lead to negative health effects especially in the male reproductive system, such as decreased sperm quality and quantity, altered sex ratio and early puberty onset. For females exposed to POPs, altered reproductive tissues and pregnancy outcomes as well as endometriosis have been reported.

## Exposure During Pregnancy

POP exposure during pregnancy is of particular concern to the developing fetus.

## Transport Across the Placenta

A study about the transfer of POPs (14 organochlorine pesticides, 7 polychlorinated biphenyls and 14 polybrominated diphenyl ethers (PBDEs)) from Spanish mothers to their unborn fetus found that POP concentrations in serum from the mother were higher than from the umbilical cord and 50 placentas. Because transfer of the POPs from mother to fetus did not correspond with passive lipid-associated diffusion, authors suggested that POPs are actively transported across the placenta.

## Gestational Weight Gain and Newborn Head Circumference

A Greek study from 2014 investigated the link between maternal weight gain during pregnancy, their PCB-exposure level and PCB level in their newborn infants, their birth weight, gestational age, and head circumference. The birth weight and head circumference of the infants was the lower, the higher POP levels during prenatal development had been, but only if mothers had either excessive or inadequate weight gain during pregnancy. No correlation between POP exposure and gestational age was found. A 2013 case-control study conducted 2009 in Indian mothers and their offspring showed prenatal exposure of two types of organochlorine pesticides (HCH, DDT and DDE) impaired the growth of the fetus, reduced the birth weight, length, head circumference and chest circumference.

## Cardiovascular Disease and Cancer

POPs are lipophilic environmental toxins. They are often found in lipoproteins of organisms. A study published in 2014 found an association between the concentration of POPs in lipoproteins and the occurrence of cardiovascular disease and various cancers in human beings. The higher the concentration of POPs found in lipoproteins, the higher the occurrence of cardiovascular disease and cancer. Highly chlorinated polychlorinated biphenyls are specifically found in high concentrations in lipoproteins. Cardiovascular disease is shown to be more associated with higher concentrations of POPs in high density lipoproteins and cancer is shown to be more associated with higher concentrations of POPs in low density lipoproteins and very low density lipoproteins.

## Obesity

There have been many recent studies assessing the connection between serum POP levels in individuals and instances of obesity. A study released in 2011 found correlations between different POPs and obesity occurrence in individuals tested. The statistically significant findings from the study show that there is actually a negative correlation between various PCB congener serum levels and obesity in individuals tested. The study also showed a positive correlation between beta-hexachlorocyclohexane and various dioxin serum levels and obesity in individuals tested. Obesity was determined using the Body Mass Index (BMI). One proposed explanation in the study is that PCBs are very lipophilic, therefore they are easily stored and captured in the fat deposits in human beings. Obese individuals have higher amounts of fat deposits in their body, and thus more PCBs could be captured in the fat deposits leading to less PCBs circulating in blood serum. The study provides evidence demonstrating that the correlation between POP serum levels and obesity occurrence is more complicated than previously expected. The same study also noted a strong positive correlation between serum POP levels and age in all individuals in the experiment.

## Diabetes

A study published in 2006 revealed a positive correlation between POP serum levels and type II diabetes in individuals, after other variables, such as age, sex, race, and socioeconomic status were adjusted for. The correlation proved stronger in younger, Mexican American, and obese individuals. Individuals exposed to low doses of POPs throughout their lifetime had a higher chance for developing diabetes than individuals exposed to high concentrations of POPs for a short amount of time.

### In Urban Areas and Indoor Environments

Traditionally it was thought that human exposure to POPs occurred primarily through food, however indoor pollution patterns that characterize certain POPs have challenged this notion. Recent studies of indoor dust and air have implicated indoor environments as a major sources for human exposure via inhalation and ingestion. Furthermore, significant indoor POP pollution must be a major route of human POP exposure, considering the modern trend in spending larger proportions of life indoor. Several studies have shown that indoor (air and dust) POP levels to exceed outdoor (air and soil) POP concentrations.

### Control and Removal in the Environment

Current studies aimed at minimizing POPs in the environment are investigating their behavior in photo catalytic oxidation reactions. POPs that are found in humans and in aquatic environments the most are the main subjects of these experiments. Aromatic and aliphatic degradation products have been identified in these reactions. Photochemical degradation is negligible compared to photocatalytic degradation. However, proper removal techniques of POPs from the environment are still unclear, due to fear that more toxic byproducts may result from uninvestigated degradation techniques. Current efforts are more focused on banning the use and production of POPs worldwide rather than removal of POPs.

## Pollution

Pollution is the introduction of contaminants into the natural environment that cause adverse change. Pollution can take the form of chemical substances or energy, such as noise, heat or light. Pollutants, the components of pollution, can be either foreign substances/energies or naturally occurring contaminants. Pollution is often classed as point source or nonpoint source pollution.

Thermal oxidizers purify industrial air flows.

The litter problem on the coast of Guyana, 2010

## History

Air pollution has always accompanied civilizations. Pollution started from prehistoric times when man created the first fires. According to a 1983 article in the journal *Science,* "soot" found on ceilings of prehistoric caves provides ample evidence of the high levels of pollution that was associated with inadequate ventilation of open fires. Metal forging appears to be a key turning point in the creation of significant air pollution levels outside the home. Core samples of glaciers in Greenland indicate increases in pollution associated with Greek, Roman and Chinese metal production, but at that time the pollution was comparatively small and could be handled by nature.

## Urban Pollution

Air pollution in the US, 1973

The burning of coal and wood, and the presence of many horses in concentrated areas made the cities the cesspools of pollution. The Industrial Revolution brought an infusion of untreated chemicals and wastes into local streams that served as the water supply. King Edward I of England banned the burning of sea-coal by proclamation in London in 1272, after its smoke became a problem. But the fuel was so common in England that this earliest of names for it was acquired because it could be carted away from some shores by the wheelbarrow.

It was the industrial revolution that gave birth to environmental pollution as we know it today. London also recorded one of the earlier extreme cases of water quality problems with the Great Stink on the Thames of 1858, which led to construction of the London sewerage system soon afterward. Pollution issues escalated as population growth far exceeded view ability of neighborhoods to handle their waste problem. Reformers began to demand sewer systems, and clean water.

In 1870, the sanitary conditions in Berlin were among the worst in Europe. August Bebel recalled conditions before a modern sewer system was built in the late 1870s:

> "Waste-water from the houses collected in the gutters running alongside the curbs and emitted a truly fearsome smell. There were no public toilets in the streets or squares. Visitors, especially women, often became desperate when nature called. In the public buildings the sanitary facilities were unbelievably primitive....As a metropolis, Berlin did not emerge from a state of barbarism into civilization until after 1870."

The primitive conditions were intolerable for a world national capital, and the Imperial German government brought in its scientists, engineers and urban planners to not only solve the deficiencies but to forge Berlin as the world's model city. A British expert in 1906 concluded that Berlin represented "the most complete application of science, order and method of public life," adding "it is a marvel of civic administration, the most modern and most perfectly organized city that there is."

The emergence of great factories and consumption of immense quantities of coal gave rise to unprecedented air pollution and the large volume of industrial chemical discharges added to the growing load of untreated human waste. Chicago and Cincinnati were the first two American cities to enact laws ensuring cleaner air in 1881. Pollution became a major issue in the United States in the early twentieth century, as progressive reformers took issue with air pollution caused by coal burning, water pollution caused by bad sanitation, and street pollution caused by the 3 million horses who worked in American cities in 1900, generating large quantities of urine and manure. As historian Martin Melosi notes, The generation that first saw automobiles replacing the horses saw cars as "miracles of cleanliness.". By the 1940s, however, automobile-caused smog was a major issue in Los Angeles.

Other cities followed around the country until early in the 20th century, when the short lived Office of Air Pollution was created under the Department of the Interior. Extreme smog events were experienced by the cities of Los Angeles and Donora, Pennsylvania in the late 1940s, serving as another public reminder. Air pollution would continue to be a problem in England, especially later during the industrial revolution, and extending into the recent past with the Great Smog of 1952.

Awareness of atmospheric pollution spread widely after World War II, with fears triggered by reports of radioactive fallout from atomic warfare and testing. Then a non-nuclear event, The Great Smog of 1952 in London, killed at least 4000 people. This prompted some of the first major modern environmental legislation, The Clean Air Act of 1956.

Pollution began to draw major public attention in the United States between the mid-1950s and early 1970s, when Congress passed the Noise Control Act, the Clean Air Act, the Clean Water Act and the National Environmental Policy Act.

Smog Pollution in Taiwan

Severe incidents of pollution helped increase consciousness. PCB dumping in the Hudson River resulted in a ban by the EPA on consumption of its fish in 1974. Long-term dioxin contamination at Love Canal starting in 1947 became a national news story in 1978 and led to the Superfund legislation of 1980. The pollution of industrial land gave rise to the name brownfield, a term now common in city planning.

The development of nuclear science introduced radioactive contamination, which can remain lethally radioactive for hundreds of thousands of years. Lake Karachay, named by the Worldwatch Institute as the "most polluted spot" on earth, served as a disposal site for the Soviet Union throughout the 1950s and 1960s. Chelyabinsk, Russia, is considered the "Most polluted place on the planet".

Nuclear weapons continued to be tested in the Cold War, especially in the earlier stages of their development. The toll on the worst-affected populations and the growth since then in understanding about the critical threat to human health posed by radioactivity has also been a prohibitive complication associated with nuclear power. Though extreme care is practiced in that industry, the potential for disaster suggested by incidents such as those at Three Mile Island and Chernobyl pose a lingering specter of public mistrust. Worldwide publicity has been intense on those disasters. Widespread support for test ban treaties has ended almost all nuclear testing in the atmosphere.

International catastrophes such as the wreck of the Amoco Cadiz oil tanker off the coast of Brittany in 1978 and the Bhopal disaster in 1984 have demonstrated the universality of such events and the scale on which efforts to address them needed to engage. The borderless nature of atmosphere and oceans inevitably resulted in the implication of pollution on a planetary level with the issue of global warming. Most recently the term persistent organic pollutant (POP) has come to describe a group of chemicals such as PBDEs and PFCs among others. Though their effects remain somewhat less well understood owing to a lack of experimental data, they have been detected in various ecological habitats far removed from industrial activity such as the Arctic, demonstrating diffusion and bioaccumulation after only a relatively brief period of widespread use.

A much more recently discovered problem is the Great Pacific Garbage Patch, a huge concentration of plastics, chemical sludge and other debris which has been collected into a large area of the Pacific Ocean by the North Pacific Gyre. This is a less well known pollution problem than the others described above, but nonetheless has multiple and serious consequences such as increasing wildlife mortality, the spread of invasive species and human ingestion of toxic chemicals. Organizations such as 5 Gyres have researched the pollution and, along with artists like Marina DeBris, are working toward publicizing the issue.

Pollution introduced by light at night is becoming a global problem, more severe in urban centres, but nonetheless contaminating also large territories, far away from towns.

Growing evidence of local and global pollution and an increasingly informed public over time have given rise to environmentalism and the environmental movement, which generally seek to limit human impact on the environment.

## Forms of Pollution

The Lachine Canal in Montreal, Quebec, Canada.

Blue drain and yellow fish symbol used by the UK Environment Agency to raise awareness of the ecological impacts of contaminating surface drainage.

The major forms of pollution are listed below along with the particular contaminant relevant to each of them:

- Air pollution: the release of chemicals and particulates into the atmosphere. Common gaseous pollutants include carbon monoxide, sulfur dioxide, chlorofluorocarbons (CFCs) and nitrogen oxides produced by industry and motor vehicles. Photochemical ozone and smog are created as nitrogen oxides and hydrocarbons react to sunlight. Particulate matter, or fine dust is characterized by their micrometre size $PM_{10}$ to $PM_{2.5}$.
- Light pollution: includes light trespass, over-illumination and astronomical interference.
- Littering: the criminal throwing of inappropriate man-made objects, unremoved, onto public and private properties.
- Noise pollution: which encompasses roadway noise, aircraft noise, industrial noise as well as high-intensity sonar.
- Soil contamination occurs when chemicals are released by spill or underground leakage. Among the most significant soil contaminants are hydrocarbons, heavy metals, MTBE, herbicides, pesticides and chlorinated hydrocarbons.
- Radioactive contamination, resulting from 20th century activities in atomic physics, such as nuclear power generation and nuclear weapons research, manufacture and deployment.
- Thermal pollution, is a temperature change in natural water bodies caused by human influence, such as use of water as coolant in a power plant.
- Visual pollution, which can refer to the presence of overhead power lines, motorway billboards, scarred landforms (as from strip mining), open storage of trash, municipal solid waste or space debris.
- Water pollution, by the discharge of wastewater from commercial and industrial waste (intentionally or through spills) into surface waters; discharges of untreated domestic sewage, and chemical contaminants, such as chlorine, from treated sewage; release of waste and contaminants into surface runoff flowing to surface waters (including urban runoff and agricultural runoff, which may contain chemical fertilizers and pesticides); waste disposal and leaching into groundwater; eutrophication and littering.
- Plastic pollution: involves the accumulation of plastic products in the environment that adversely affects wildlife, wildlife habitat, or humans.

## Pollutants

A pollutant is a waste material that pollutes air, water or soil. Three factors determine the severity of a pollutant: its chemical nature, the concentration and the persistence.

## Cost of Pollution

Pollution has cost. Manufacturing activities that cause air pollution impose health and clean-up costs on the whole society, whereas the neighbors of an individual who chooses to fire-proof his home may benefit from a reduced risk of a fire spreading to their own houses. If external costs exist, such as pollution, the producer may choose to produce more of the product than would be produced if the producer were required to pay all associated environmental costs. Because responsibility or consequence for self-directed action lies partly outside the self, an element of externalization is involved. If there are external benefits, such as in public safety, less of the good may be produced than would be the case if the producer were to receive payment for the external benefits to others.

## Sources and Causes

Air pollution produced by ships may alter clouds, affecting global temperatures.

Air pollution comes from both natural and human-made (anthropogenic) sources. However, globally human-made pollutants from combustion, construction, mining, agriculture and warfare are increasingly significant in the air pollution equation.

Motor vehicle emissions are one of the leading causes of air pollution. China, United States, Russia, India Mexico, and Japan are the world leaders in air pollution emissions. Principal stationary pollution sources include chemical plants, coal-fired power plants, oil refineries, petrochemical plants, nuclear waste disposal activity, incinerators, large livestock farms (dairy cows, pigs, poultry, etc.), PVC factories, metals production factories, plastics factories, and other heavy industry. Agricultural air pollution comes from contemporary practices which include clear felling and burning of natural vegetation as well as spraying of pesticides and herbicides

About 400 million metric tons of hazardous wastes are generated each year. The United States alone produces about 250 million metric tons. Americans constitute less than 5% of the world's population, but produce roughly 25% of the world's $CO_2$, and generate approximately 30% of world's waste. In 2007, China has overtaken the United States as the world's biggest producer of $CO_2$, while still far behind based on per capita pollution - ranked 78th among the world's nations.

An industrial area, with a power plant, south of Yangzhou's downtown, China

In February 2007, a report by the Intergovernmental Panel on Climate Change (IPCC), representing the work of 2,500 scientists, economists, and policymakers from more than 120 countries, said that humans have been the primary cause of global warming since 1950. Humans have ways to cut greenhouse gas emissions and avoid the consequences of global warming, a major climate report concluded. But to change the climate, the transition from fossil fuels like coal and oil needs to occur within decades, according to the final report this year from the UN's Intergovernmental Panel on Climate Change (IPCC).

Some of the more common soil contaminants are chlorinated hydrocarbons (CFH), heavy metals (such as chromium, cadmium–found in rechargeable batteries, and lead–found in lead paint, aviation fuel and still in some countries, gasoline), MTBE, zinc, arsenic and benzene. In 2001 a series of press reports culminating in a book called *Fateful Harvest* unveiled a widespread practice of recycling industrial byproducts into fertilizer, resulting in the contamination of the soil with various metals. Ordinary municipal landfills are the source of many chemical substances entering the soil environment (and often groundwater), emanating from the wide variety of refuse accepted, especially substances illegally discarded there, or from pre-1970 landfills that may have been subject to little control in the U.S. or EU. There have also been some unusual releases of polychlorinated dibenzodioxins, commonly called *dioxins* for simplicity, such as TCDD.

Pollution can also be the consequence of a natural disaster. For example, hurricanes often involve water contamination from sewage, and petrochemical spills from ruptured boats or automobiles. Larger scale and environmental damage is not uncommon when coastal oil rigs or refineries are involved. Some sources of pollution, such as nuclear power plants or oil tankers, can produce widespread and potentially hazardous releases when accidents occur.

In the case of noise pollution the dominant source class is the motor vehicle, producing about ninety percent of all unwanted noise worldwide.

## Effects

### Human Health

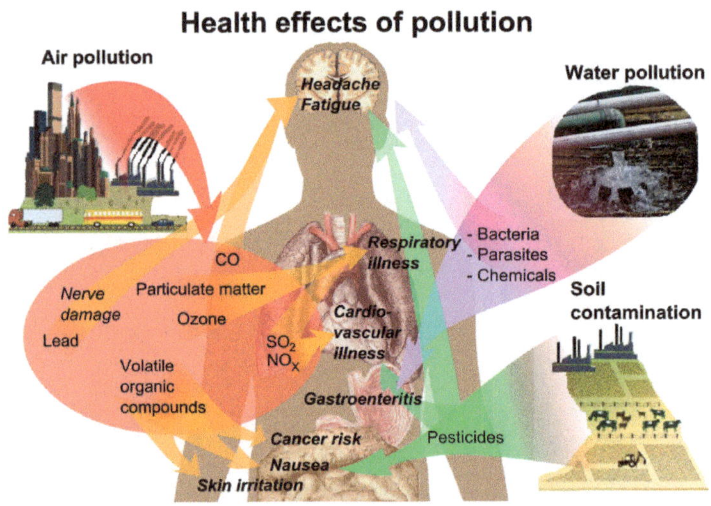


Overview of main health effects on humans from some common types of pollution.

Adverse air quality can kill many organisms including humans. Ozone pollution can cause respiratory disease, cardiovascular disease, throat inflammation, chest pain, and congestion. Water pollution causes approximately 14,000 deaths per day, mostly due to contamination of drinking water by untreated sewage in developing countries. An estimated 500 million Indians have no access to a proper toilet, Over ten million people in India fell ill with waterborne illnesses in 2013, and 1,535 people died, most of them children. Nearly 500 million Chinese lack access to safe drinking water. A 2010 analysis estimated that 1.2 million people died prematurely each year in China because of air pollution. The WHO estimated in 2007 that air pollution causes half a million deaths per year in India. Studies have estimated that the number of people killed annually in the United States could be over 50,000.

Oil spills can cause skin irritations and rashes. Noise pollution induces hearing loss, high blood pressure, stress, and sleep disturbance. Mercury has been linked to developmental deficits in children and neurologic symptoms. Older people are majorly exposed to diseases induced by air pollution. Those with heart or lung disorders are at additional risk. Children and infants are also at serious risk. Lead and other heavy metals have been shown to cause neurological problems. Chemical and radioactive substances can cause cancer and as well as birth defects.

## Environment

Pollution has been found to be present widely in the environment. There are a number of effects of this:

- Biomagnification describes situations where toxins (such as heavy metals) may pass through trophic levels, becoming exponentially more concentrated in the process.
- Carbon dioxide emissions cause ocean acidification, the ongoing decrease in the pH of the Earth's oceans as $CO_2$ becomes dissolved.
- The emission of greenhouse gases leads to global warming which affects ecosystems in many ways.
- Invasive species can out compete native species and reduce biodiversity. Invasive plants can contribute debris and biomolecules (allelopathy) that can alter soil and chemical compositions of an environment, often reducing native species competitiveness.
- Nitrogen oxides are removed from the air by rain and fertilise land which can change the species composition of ecosystems.
- Smog and haze can reduce the amount of sunlight received by plants to carry out photosynthesis and leads to the production of tropospheric ozone which damages plants.
- Soil can become infertile and unsuitable for plants. This will affect other organisms in the food web.
- Sulfur dioxide and nitrogen oxides can cause acid rain which lowers the pH value of soil.

## Environmental Health Information

The Toxicology and Environmental Health Information Program (TEHIP) at the United States National Library of Medicine (NLM) maintains a comprehensive toxicology and environmental health web site that includes access to resources produced by TEHIP and by other government agencies and organizations. This web site includes links to databases, bibliographies, tutorials, and other scientific and consumer-oriented resources. TEHIP also is responsible for the Toxicology Data Network (TOXNET) an integrated system of toxicology and environmental health databases that are available free of charge on the web.

TOXMAP is a Geographic Information System (GIS) that is part of TOXNET. TOXMAP uses maps of the United States to help users visually explore data from the United States Environmental Protection Agency's (EPA) Toxics Release Inventory and Superfund Basic Research Programs.

## Worker Productivity

A number of studies show that pollution has an adverse effect on the productivity of both indoor and outdoor workers.

## Regulation and Monitoring

To protect the environment from the adverse effects of pollution, many nations worldwide have enacted legislation to regulate various types of pollution as well as to mitigate the adverse effects of pollution.

## Pollution Control

A litter trap catches floating waste in the Yarra River, east-central Victoria, Australia

Air pollution control system, known as a Thermal oxidizer, decomposes hazard gases from industrial air streams at a factory in the United States of America.

Gas nozzle with vapor recovery

Pollution control is a term used in environmental management. It means the control of emissions and effluents into air, water or soil. Without pollution control, the waste products from overconsumption, heating, agriculture, mining, manufacturing, transportation and other human activities, whether they accumulate or disperse, will degrade the environment. In the hierarchy of controls, pollution prevention and waste minimization are more desirable than pollution control. In the field of land development, low impact development is a similar technique for the prevention of urban runoff.

## Practices

- Recycling
- Reusing
- Waste minimisation
- Mitigating
- Preventing
- Compost

## Pollution Control Devices

- Air pollution control
    - Thermal oxidizer
- Dust collection systems
    - Baghouses
    - Cyclones
    - Electrostatic precipitators
- Scrubbers
    - Baffle spray scrubber
    - Cyclonic spray scrubber
    - Ejector venturi scrubber

    - Mechanically aided scrubber
    - Spray tower
    - Wet scrubber
- Sewage treatment
    - Sedimentation (Primary treatment)
    - Activated sludge biotreaters (Secondary treatment; also used for industrial wastewater)
    - Aerated lagoons
    - Constructed wetlands (also used for urban runoff)
- Industrial wastewater treatment
    - API oil-water separators
    - Biofilters
    - Dissolved air flotation (DAF)
    - Powdered activated carbon treatment
    - Ultrafiltration
- Vapor recovery systems
- Phytoremediation

## Perspectives

The earliest precursor of pollution generated by life forms would have been a natural function of their existence. The attendant consequences on viability and population levels fell within the sphere of natural selection. These would have included the demise of a population locally or ultimately, species extinction. Processes that were untenable would have resulted in a new balance brought about by changes and adaptations. At the extremes, for any form of life, consideration of pollution is superseded by that of survival.

For humankind, the factor of technology is a distinguishing and critical consideration, both as an enabler and an additional source of byproducts. Short of survival, human concerns include the range from quality of life to health hazards. Since science holds experimental demonstration to be definitive, modern treatment of toxicity or environmental harm involves defining a level at which an effect is observable. Common examples of fields where practical measurement is crucial include automobile emissions control, industrial exposure (e.g. Occupational Safety and Health Administration (OSHA) PELs), toxicology (e.g. $LD_{50}$), and medicine (e.g. medication and radiation doses).

"The solution to pollution is dilution", is a dictum which summarizes a traditional approach to pollution management whereby sufficiently diluted pollution is not harmful. It is well-suited to some other modern, locally scoped applications such as laboratory safety procedure and hazardous material release emergency management. But it assumes that the dilutant is in virtually unlimited supply for the application or that resulting dilutions are acceptable in all cases.

Such simple treatment for environmental pollution on a wider scale might have had greater merit in earlier centuries when physical survival was often the highest imperative, human population and densities were lower, technologies were simpler and their byproducts more benign. But these are often no longer the case. Furthermore, advances have enabled measurement of concentrations not possible before. The use of statistical methods in evaluating outcomes has given currency to the principle of probable harm in cases where assessment is warranted but resorting to deterministic models is impractical or infeasible. In addition, consideration of the environment beyond direct impact on human beings has gained prominence.

Yet in the absence of a superseding principle, this older approach predominates practices throughout the world. It is the basis by which to gauge concentrations of effluent for legal release, exceeding which penalties are assessed or restrictions applied. One such superseding principle is contained in modern hazardous waste laws in developed countries, as the process of diluting hazardous waste to make it non-hazardous is usually a regulated treatment process. Migration from pollution dilution to elimination in many cases can be confronted by challenging economical and technological barriers.

## Greenhouse Gases and Global Warming

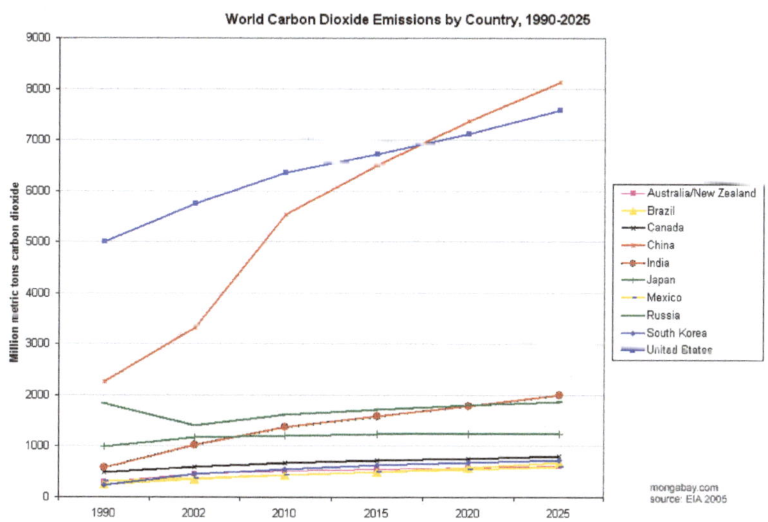


Historical and projected $CO_2$ emissions by country (as of 2005). Source: Energy Information Administration.

Carbon dioxide, while vital for photosynthesis, is sometimes referred to as pollution, because raised levels of the gas in the atmosphere are affecting the Earth's climate. Disruption of the environment can also highlight the connection between areas of pollution that would normally be classified separately, such as those of water and air. Recent studies have investigated the potential for long-term rising levels of atmospheric carbon dioxide to cause slight but critical increases in the acidity of ocean waters, and the possible effects of this on marine ecosystems.

## Most Polluting Industries

The Pure Earth, an international non-for-profit organization dedicated to eliminating life-threatening pollution in the developing world, issues an annual list of some of the world's most polluting industries.

- Lead-Acid Battery Recycling
- Industrial Mining and Ore Processing
- Lead Smelting
- Tannery Operations
- Artisanal Small-Scale Gold Mining
- Industrial/Municipal Dumpsites
- Industrial Estates
- Chemical Manufacturing
- Product Manufacturing
- Dye Industry

## World's Worst Polluted Places

The Pure Earth issues an annual list of some of the world's worst polluted places.

- Agbogbloshie, Ghana
- Chernobyl, Ukraine
- Citarum River, Indonesia
- Dzershinsk, Russia
- Hazaribagh, Bangladesh
- Kabwe, Zambia
- Kalimantan, Indonesia
- Matanza Riachuelo, Argentina
- Niger River Delta, Nigeria
- Norilsk, Russia

## Point Source Pollution

A point source of pollution is a single identifiable source of air, water, thermal, noise or light pollution. A point source has negligible extent, distinguishing it from other pollution source geometries. The sources are called *point sources* because in mathematical modeling, they can be approximated as a mathematical point to simplify analysis. Pollution point sources are identical to other physics, engineering, optics, and chemistry point sources and include:

Industrial point sources of air pollution.

- Air pollution from an industrial source, rather than an airport or a road, considered a line source or a forest fire which is considered an area source, or volume source.
- Water pollution from an oil refinery wastewater discharge outlet
- Noise pollution from a jet engine
- Disruptive seismic vibration from a localized seismic study
- Light pollution from an intrusive street light
- Thermal pollution from an industrial process outfall
- Radio emissions from an interference-producing electrical device

## Nonpoint Source Pollution

Muddy river

Nonpoint source (NPS) pollution is a term used to describe pollution resulting from many diffuse sources, in direct contrast to point source pollution which results from a single source. Nonpoint source pollution generally results from land runoff, precipitation, atmospheric deposition, drainage, seepage, or hydrological modification (rainfall or snowmelt) where tracing the pollution back to a single source is difficult.

Non-point source water pollution affects a water body from sources such as polluted runoff from agricultural areas draining into a river, or wind-borne debris blowing out to sea. Non-point source air pollution affects air quality from sources such as smokestacks or car tailpipes. Although these pollutants have originated from a point source, the long-range transport ability and multiple sources of the pollutant make it a non-point source of pollution. Non-point source pollution can be contrasted with point source pollution, where discharges occur to a body of water or into the atmosphere at a single location.

NPS may derive from many different sources with no specific solution may change to rectify the problem, making it difficult to regulate. Non point source water pollution is difficult to control because it comes from the everyday activities of many different people, such as fertilizing a lawn, using a pesticide, or constructing a road or building.

It is the leading cause of water pollution in the United States today, with polluted runoff from agriculture the primary cause.

Other significant sources of runoff include hydrological and habitat modification, and silviculture (forestry).

Contaminated stormwater washed off parking lots, roads and highways, and lawns (often containing fertilizers and pesticides) is called urban runoff. This runoff is often classified as a type of NPS pollution. Some people may also consider it a point source because many times it is channeled into municipal storm drain systems and discharged through pipes to nearby surface waters. However, not all urban runoff flows through storm drain systems before entering water bodies. Some may flow directly into water bodies, especially in developing and suburban areas. Also, unlike other types of point sources, such as industrial discharges, sewage treatment plants and other operations, pollution in urban runoff cannot be attributed to one activity or even group of activities. Therefore, because it is not caused by an easily identified and regulated activity, urban runoff pollution sources are also often treated as true non-point sources as municipalities work to abate them.

## Principal Types

### Sediment

Sediment (loose soil) includes silt (fine particles) and suspended solids (larger particles). Sediment may enter surface waters from eroding stream banks, and from surface runoff due to improper plant cover on urban and rural land. Sediment creates turbidity (cloudiness) in water bodies, reducing the amount of light reaching lower depths, which can inhibit growth of submerged aquatic plants and consequently affect species which are dependent on them, such as fish and shellfish. High turbidity levels also inhibit drinking water purification systems.

Sediment can also be discharged from multiple different sources. Sources include construction sites (although these are point sources, which can be managed with erosion controls and sediment controls), agricultural fields, stream banks, and highly disturbed areas.

### Nutrients

Nutrients mainly refers to inorganic matter from runoff, landfills, livestock operations and crop lands. The two primary nutrients of concern are phosphorus and nitrogen.

Phosphorus is a nutrient that occurs in many forms that are bioavailable. It is notoriously over-abundant in human sewage sludge. It is a main ingredient in many fertilizers used for agriculture as well as on residential and commercial properties, and may become a limiting nutrient in freshwater systems and some estuaries. Phosphorus is most often transported to water bodies via soil erosion because many forms of phosphorus tend to be adsorbed on to soil particles. Excess amounts of phosphorus in aquatic systems (particularly freshwater lakes, reservoirs, and ponds) leads to proliferation of microscopic algae called phytoplankton. The increase of organic matter supply due to the excessive growth of the phytoplankton is called eutrophication. A common symptom of eutrophication is algae blooms that can produce unsightly surface scums, shade out beneficial types of plants, and poison the water due to toxins produced by the algae. These toxins are a particular problem in systems used for drinking water because some toxins can cause human illness and removal of the toxins is difficult and expensive. Bacterial decomposition of algal blooms consumes dissolved oxygen in the water, generating hypoxia with detrimental consequences for fish and aquatic invertebrates.

Nitrogen is the other key ingredient in fertilizers, and it generally becomes a pollutant in saltwater or brackish estuarine systems where nitrogen is a limiting nutrient. Similar to phosphorus in fresh-waters, excess amounts of bioavailable nitrogen in marine systems lead to eutrophication and algae blooms. Hypoxia is an increasingly common result of eutrophication in marine systems and can impact large areas of estuaries, bays, and near shore coastal waters. Each summer, hypoxic conditions form in bottom waters where the Mississippi River enters the Gulf of Mexico. During recent summers, the aereal extent of this "dead zone" is comparable to the area of New Jersey and has major detrimental consequences for fisheries in the region.

Nitrogen is most often transported by water as nitrate ($NO_3$). The nitrogen is usually added to a watershed as organic-N or ammonia ($NH_3$), so nitrogen stays attached to the soil until oxidation converts it into nitrate. Since the nitrate is generally already incorporated into the soil, the water traveling through the soil (i.e., interflow and tile drainage) is the most likely to transport it, rather than surface runoff.

## Toxic Contaminants and Chemicals

Compounds including heavy metals like lead, mercury, zinc, and cadmium, organics like polychlorinated biphenyls (PCBs) and polycyclic aromatic hydrocarbons (PAHs), fire retardants, and other substances are resistant to breakdown. These contaminants can come from a variety of sources including human sewage sludge, mining operations, vehicle emissions, fossil fuel combustion, urban runoff, industrial operations and landfills.

Toxic chemicals mainly include organic compounds and inorganic compounds. These compounds include pesticides like DDT, acids, and salts that have severe effects to the ecosystem and water-bodies. These compounds can threaten the health of both humans and aquatic species while being resistant to environmental breakdown, thus allowing them to persist in the environment. These toxic chemicals could come from croplands, nurseries, orchards, building sites, gardens, lawns and landfills.

Acids and salts mainly are inorganic pollutants from irrigated lands, mining operations, urban runoff, industrial sites and landfills.

## Pathogens

Pathogens are bacteria and viruses that can be found in water and cause diseases in humans. Typically, pathogens cause disease when they are present in public drinking water supplies. Pathogens found in contaminated runoff may include:

- *Cryptosporidium parvum*
- *Giardia lamblia*
- *Salmonella*
- *Novovirus* and other viruses
- Parasitic worms (helminths).

Coliform bacteria and fecal matter may also be detected in runoff. These bacteria are a commonly used indicator of water pollution, but not an actual cause of disease.

Pathogens may contaminate runoff due to poorly managed livestock operations, faulty septic systems, improper handling of pet waste, the over application of human sewage sludge, contaminated storm sewers, and sanitary sewer overflows.

## Principal Sources

### Urban and Suburban Areas

Urban and suburban areas are a main sources of nonpoint source pollution due to the amount of runoff that is produced due to the large amount of paved surfaces. Paved surfaces, such as asphalt and concrete are impervious to water penetrating them. Any water that is on contact with these surfaces will run off and be absorbed by the surrounding environment. These surfaces make it easier for stormwater to carry pollutants into the surrounding soil.

Construction sites tend to have disturbed soil that is easily eroded by precipitation like rain, snow, and hail. Additionally, discarded debris on the site can be carried away by runoff waters and enter the aquatic environment.

Typically, in suburban areas, chemicals are used for lawn care. These chemicals can end up in runoff and enter the surrounding environment via storm drains in the city. Since the water in storm drains is not treated before flowing into surrounding water bodies, the chemicals enter the water directly.

### Agricultural Operations

Agricultural operations account for a large percentage of all nonpoint source pollution in the United States. When large tracts of land are plowed to grow crops, it exposes and loosens soil that was once buried. This makes the exposed soil more vulnerable to erosion during rainstorms. It also can increase the amount of fertilizer and pesticides carried into nearby bodies of water.

### Atmospheric Inputs

Atmospheric inputs of pollution into the air can come from multiple sources. Typically, industrial facilities, like factories, emit air pollution via a smokestack. Although this is a point source, due to the distributional nature, long-range transport, and multiple sources of the pollution, it is considered a nonpoint source. Additionally, atmospheric pollution can become water pollution, by being washed out of the atmosphere in the form of rain or snow.

## Highway Runoff

Highway runoff accounts for a small but widespread percentage of all nonpoint source pollution. Harned (1988) estimated that runoff loads were composed of atmospheric fallout (9%), vehicle deposition (25%) and highway maintenance materials (67%) he also estimated that about 9 percent of these loads were reentrained in the atmosphere.

## Forestry and Mining Operations

Forestry and mining operations can have significant inputs to non-point source pollution.

## Forestry

Forestry operations reduce the number of trees in a given area, thus reducing the oxygen levels in that area as well. This action, coupled with the heavy machinery rolling over the soil increases the risk of erosion.

## Mining

Active mining operations are considered point sources, however runoff from abandoned mining operations contribute to nonpoint source pollution. In strip mining operations, the top of the mountain is removed to expose the desired ore. If this area is not properly reclaimed once the mining has finished, soil erosion can occur. Additionally, there can be chemical reactions with the air and newly exposed rock to create acidic runoff. Water that seeps out of abandoned subsurface mines can also be highly acidic. This can seep into the nearest body of water and change the pH in the aquatic environment.

## Marinas and Boating Activities

Chemicals used for boat maintenance, like paint, solvents, and oils find their way into water through runoff. Additionally, spilling fuels or leaking fuels directly into the water from boats contribute to nonpoint source pollution. Nutrient and bacteria levels are increased by poorly maintained sanitary waste receptacles on the boat and pump-out stations.

## Control

Contour buffer strips used to retain soil and reduce erosion

## Urban and Suburban Areas

To control non-point source pollution, many different approaches can be undertaken in both urban and suburban areas. Buffer strips provide a barrier of grass in between impervious paving material like parking lots and roads, and the closest body of water. This allows the soil to absorb any pollution before it enters the local aquatic system. Retention ponds can be built in drainage areas to create an aquatic buffer between runoff pollution and the aquatic environment. Runoff and storm water drain into the retention pond allowing for the contaminants to settle out and become trapped in the pond. The use of porous pavement allows for rain and storm water to drain into the ground beneath the pavement, reducing the amount of runoff that drains directly into the water body. Restoration methods such as constructing wetlands are also used to slow runoff as well as absorb contamination.

Construction sites typically implement simple measures to reduce pollution and runoff. Firstly, sediment or silt fences are erected around construction sites to reduce the amount of sediment and large material draining into the nearby water body. Secondly, laying grass or straw along the border of construction sites also work to reduce nonpoint source pollution.

In areas served by single-home septic systems, local government regulations can force septic system maintenance to ensure compliance with water quality standards. In Washington (state), a novel approach was developed through a creation of a "shellfish protection district" when either a commercial or recreational shellfish bed is downgraded because of ongoing nonpoint source pollution. The shellfish protection district is a geographic area designated by a county to protect water quality and tideland resources, and provides a mechanism to generate local funds for water quality services to control nonpoint sources of pollution. At least two shellfish protection districts in south Puget Sound have instituted septic system operation and maintenance requirements with program fees tied directly to property taxes.

## Agricultural Operations

To control sediment and runoff, farmers may utilize erosion controls to reduce runoff flows and retain soil on their fields. Common techniques include contour plowing, crop mulching, crop rotation, planting perennial crops and installing riparian buffers. *Conservation tillage* is a concept used to reduce runoff while planting a new crop. The farmer leaves some crop reside from the previous planting in the ground to help prevent runoff during the planting process.

Nutrients are typically applied to farmland as commercial fertilizer; animal manure; or spraying of municipal or industrial wastewater (effluent) or sludge. Nutrients may also enter runoff from crop residues, irrigation water, wildlife, and atmospheric deposition. Farmers can develop and implement nutrient management plans to reduce excess application of nutrients.

To minimize pesticide impacts, farmers may use Integrated Pest Management (IPM) techniques (which can include biological pest control) to maintain control over pests, reduce reliance on chemical pesticides, and protect water quality.

## Forestry Operations

With a well-planned placement of both logging trails, also called skid trails, can reduce the amount of sediment generated. By planning the trails location as far away from the logging activity as possible as well as contouring the trails with the land, it can reduce the amount of loose sediment in the runoff. Additionally, by replanting trees on the land after logging, it provides a structure for the soil to regain stability as well as replaces the logged environment.

## Marinas

Installing shut off valves on fuel pumps at a marina dock can help reduce the amount of spillover into the water. Additionally, pump-out stations that are easily accessible to boaters in a marina can provide a clean place in which to dispose of sanitary waste without dumping it directly into the water. Finally, something as simple as having trash containers around a marina can prevent larger objects entering the water.

## References


- Beychok, Milton R. (1967). Aqueous Wastes from Petroleum and Petrochemical Plants (1st ed.). John Wiley & Sons. ISBN 0-471-07189-7. LCCN 67019834
- Ritter L; Solomon KR; Forget J; Stemeroff M; O'Leary C. "Persistent organic pollutants" (PDF). United Nations Environment Programme. Retrieved 2007-09-16
- El-Shahawi, M.S., Hamza, A., Bashammakhb, A.S., Al-Saggaf, W.T. (2010). An overview on the accumulation, distribution, transformations, toxicity and analytical methods for the monitoring of persistent organic pollutants. Talanta. 80, 1587–1597. doi:10.1016/j.talanta.2009.09.055
- Astoviza, Malena J. (15 April 2014). "Evaluación de la distribución de contaminantes orgánicos persistentes (COPs) en aire en la zona de la cuenca del Plata mediante muestreadores pasivos artificiales" (in Spanish): 160. Retrieved 16 April 2014
- Kelly, B.C., Ikonomou, M.G., Blair, J.D., Morin, A.E., Gobas, F.A.P.C. (2007). Food Web-Specific Biomagnification of Persistent Organic Pollutants. Science. 317, 236–239. doi:10.1126/science.1138275
- Wania, F., Mackay, D. (1996). Tracking the Distribution of Persistent Organic Pollutants. Environmental Science & Technology. 30 (9), 390A–396A. doi:10.1021/es962399q
- Fleming, James R.; Knorr, Bethany R. "History of the Clean Air Act". American Meteorological Society. Retrieved 2006-02-14
- Yu, G.W.,, Laseter, J., Mylander, C. Persistent organic pollutants in serum and several different fat compartments in humans. J Environ Public Health. 2011;2011:417980. doi:10.1155/2011/417980. PMID 21647350
- Lohmanna, R., Breivikb, K., Dachsd, J., Muire, D. (2007). Global fate of POPs: Current and future research directions. 150, 150–165. doi:10.1016/j.envpol.2007.06.051
- World Resources Institute: August 2008 Monthly Update: Air Pollution's Causes, Consequences and Solutions Submitted by Matt Kallman on Wed, 2008-08-20 18:22. Retrieved on April 17, 2009
- Szabo DT, Loccisano AE (March 30, 2012). A. Schecter, ed. "POPs and Human Health Risk Assessment". Dioxins and Persistent Organic Pollutants. John Wiley & Sons. 3rd. doi:10.1002/9781118184141.ch19
- Waterhealthconnection.org Overview of Waterborne Disease Trends By Patricia L. Meinhardt, MD, MPH, MA, Author. Retrieved on April 16, 2009
- Vizcaino, E; Grimalt JO; Fernández-Somoano A; Tardon A (2014). "Transport of persistent organic pollutants across the human placenta". Environ Int. 65: 107–115. doi:10.1016/j.envint.2014.01.004. PMID 24486968
- Dewan, Jain V,; Gupta P; Banerjee BD. (February 2013). "Organochlorine pesticide residues in maternal blood, cord blood, placenta, and breastmilk and their relation to birth size". Chemosphere. 90 (5): 1704–1710. doi:10.1016/j.chemosphere.2012.09.083. PMID 23141556

- The World's Most Polluted Places: The Top Ten of the Dirty Thirty (PDF), archived from the original (PDF) on 2007-10-11, retrieved 2013-12-10
- Ljunggren SA, Helmfrid I, Salihovic S, van Bavel B, Wingren G, Lindahl M, Karlsson H. Persistent organic pollutants distribution in lipoprotein fractions in relation to cardiovascular disease and cancer. Environ Int. 2014;65:93–99. doi:10.1016/j.envint.2013.12.017. PMID 24472825
- Van Leeuwen, C.J. (2010). Risk Assessment of Chemicals: An Introduction, 2nd Ed. Dordrecht, The Netherlands: Springer. ISBN 978-1-4020-6101-1
- Beyer, A., Mackay, D., Matthies, M., Wania, F., Webster, E. (2000). Assessing Long-Range Transport Potential of Persistent Organic Pollutants. Environmental Sciences & Technology. 34(4), 699–703. doi:10.1021/es990207w

# 2

# Air, Noise and Water Pollution and Its Effects

Pollution can damage the environment immensely. There are various types of pollutions such as noise pollution, water pollution, air pollution, light pollution and toxic hotspot. The increase in pollution has resulted in the destruction of habitats and extinction of several species. Air, noise and water pollution can be best understood in confluence with the major topics listed in the following chapter.

## Air Pollution

Air pollution from a coking oven

Air pollution occurs when harmful substances including particulates and biological molecules are introduced into Earth's atmosphere. It may cause diseases, allergies or death in humans; it may also cause harm to other living organisms such as animals and food crops, and may damage the natural or built environment. Human activity and natural processes can both generate air pollution.

Indoor air pollution and poor urban air quality are listed as two of the world's worst toxic pollution problems in the 2008 Blacksmith Institute World's Worst Polluted Places report. According to the 2014 WHO report, air pollution in 2012 caused the deaths of around 7 million people worldwide, an estimate roughly matched by the International Energy Agency.

### Pollutants

An air pollutant is a substance in the air that can have adverse effects on humans and the ecosystem. The substance can be solid particles, liquid droplets, or gases. A pollutant can be of natural origin or man-made. Pollutants are classified as primary or secondary. Primary pollutants are usually produced from a process, such as ash from a volcanic eruption. Other examples include carbon monoxide gas from motor vehicle exhaust, or the sulfur dioxide released from factories.

Secondary pollutants are not emitted directly. Rather, they form in the air when primary pollutants react or interact. Ground level ozone is a prominent example of a secondary pollutant. Some pollutants may be both primary and secondary: they are both emitted directly and formed from other primary pollutants.

Before flue-gas desulfurization was installed, the emissions from this power plant in New Mexico contained excessive amounts of sulfur dioxide.

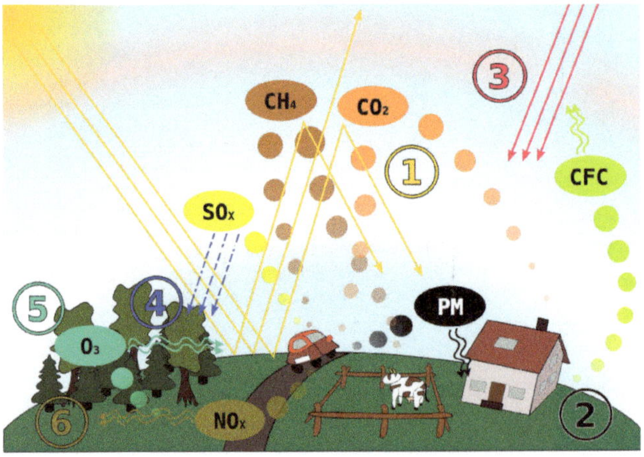


Schematic drawing, causes and effects of air pollution: (1) greenhouse effect, (2) particulate contamination, (3) increased UV radiation, (4) acid rain, (5) increased ground level ozone concentration, (6) increased levels of nitrogen oxides.

Substances emitted into the atmosphere by human activity include:

- Carbon dioxide ($CO_2$) - Because of its role as a greenhouse gas it has been described as "the leading pollutant" and "the worst climate pollution". Carbon dioxide is a natural component of the atmosphere, essential for plant life and given off by the human respiratory system. This question of terminology has practical effects, for example as determining whether the U.S. Clean Air Act is deemed to regulate $CO_2$ emissions. $CO_2$ currently forms about 405 parts per million (ppm) of earth's atmosphere, compared to about 280 ppm in pre-industrial times, and billions of metric tons of $CO_2$ are emitted annually by burning of fossil fuels. $CO_2$ increase in earth's atmosphere has been accelerating.
- Sulfur oxides ($SO_x$) - particularly sulfur dioxide, a chemical compound with the formula $SO_2$. $SO_2$ is produced by volcanoes and in various industrial processes. Coal and petroleum

often contain sulfur compounds, and their combustion generates sulfur dioxide. Further oxidation of $SO_2$, usually in the presence of a catalyst such as $NO_2$, forms $H_2SO_4$, and thus acid rain. This is one of the causes for concern over the environmental impact of the use of these fuels as power sources.

- Nitrogen oxides ($NO_x$) - Nitrogen oxides, particularly nitrogen dioxide, are expelled from high temperature combustion, and are also produced during thunderstorms by electric discharge. They can be seen as a brown haze dome above or a plume downwind of cities. Nitrogen dioxide is a chemical compound with the formula $NO_2$. It is one of several nitrogen oxides. One of the most prominent air pollutants, this reddish-brown toxic gas has a characteristic sharp, biting odor.

- Carbon monoxide (CO) - CO is a colorless, odorless, toxic yet non-irritating gas. It is a product of incomplete combustion of fuel such as natural gas, coal or wood. Vehicular exhaust is a major source of carbon monoxide.

- Volatile organic compounds (VOC) - VOCs are a well-known outdoor air pollutant. They are categorized as either methane ($CH_4$) or non-methane (NMVOCs). Methane is an extremely efficient greenhouse gas which contributes to enhanced global warming. Other hydrocarbon VOCs are also significant greenhouse gases because of their role in creating ozone and prolonging the life of methane in the atmosphere. This effect varies depending on local air quality. The aromatic NMVOCs benzene, toluene and xylene are suspected carcinogens and may lead to leukemia with prolonged exposure. 1,3-butadiene is another dangerous compound often associated with industrial use.

- Particulates, alternatively referred to as particulate matter (PM), atmospheric particulate matter, or fine particles, are tiny particles of solid or liquid suspended in a gas. In contrast, aerosol refers to combined particles and gas. Some particulates occur naturally, originating from volcanoes, dust storms, forest and grassland fires, living vegetation, and sea spray. Human activities, such as the burning of fossil fuels in vehicles, power plants and various industrial processes also generate significant amounts of aerosols. Averaged worldwide, anthropogenic aerosols—those made by human activities—currently account for approximately 10 percent of our atmosphere. Increased levels of fine particles in the air are linked to health hazards such as heart disease, altered lung function and lung cancer.

- Persistent free radicals connected to airborne fine particles are linked to cardiopulmonary disease.

- Toxic metals, such as lead and mercury, especially their compounds.

- Chlorofluorocarbons (CFCs) - harmful to the ozone layer; emitted from products are currently banned from use. These are gases which are released from air conditioners, refrigerators, aerosol sprays, etc. On release into the air, CFCs rise to the stratosphere. Here they come in contact with other gases and damage the ozone layer. This allows harmful ultraviolet rays to reach the earth's surface. This can lead to skin cancer, eye disease and can even cause damage to plants.

- Ammonia ($NH_3$) - emitted from agricultural processes. Ammonia is a compound with the formula $NH_3$. It is normally encountered as a gas with a characteristic pungent odor. Ammonia contributes significantly to the nutritional needs of terrestrial organisms by serving as a precursor to foodstuffs and fertilizers. Ammonia, either directly or indirectly, is also a building block for the synthesis of many pharmaceuticals. Although in wide use, ammonia is both caustic and hazardous. In the atmosphere, ammonia reacts with oxides of nitrogen and sulfur to form secondary particles.
- Odours — such as from garbage, sewage, and industrial processes
- Radioactive pollutants - produced by nuclear explosions, nuclear events, war explosives, and natural processes such as the radioactive decay of radon.

Secondary pollutants include:

- Particulates created from gaseous primary pollutants and compounds in photochemical smog. Smog is a kind of air pollution. Classic smog results from large amounts of coal burning in an area caused by a mixture of smoke and sulfur dioxide. Modern smog does not usually come from coal but from vehicular and industrial emissions that are acted on in the atmosphere by ultraviolet light from the sun to form secondary pollutants that also combine with the primary emissions to form photochemical smog.
- Ground level ozone ($O_3$) formed from $NO_x$ and VOCs. Ozone ($O_3$) is a key constituent of the troposphere. It is also an important constituent of certain regions of the stratosphere commonly known as the Ozone layer. Photochemical and chemical reactions involving it drive many of the chemical processes that occur in the atmosphere by day and by night. At abnormally high concentrations brought about by human activities (largely the combustion of fossil fuel), it is a pollutant, and a constituent of smog.
- Peroxyacetyl nitrate ($C_2H_3NO_5$) - similarly formed from $NO_x$ and VOCs.

Minor air pollutants include:

- A large number of minor hazardous air pollutants. Some of these are regulated in USA under the Clean Air Act and in Europe under the Air Framework Directive
- A variety of persistent organic pollutants, which can attach to particulates

Persistent organic pollutants (POPs) are organic compounds that are resistant to environmental degradation through chemical, biological, and photolytic processes. Because of this, they have been observed to persist in the environment, to be capable of long-range transport, bioaccumulate in human and animal tissue, biomagnify in food chains, and to have potentially significant impacts on human health and the environment.

## Sources

There are various locations, activities or factors which are responsible for releasing pollutants into the atmosphere. These sources can be classified into two major categories.

Anthropogenic (man-made) sources:

Controlled burning of a field outside of Statesboro, Georgia in preparation for spring planting.

These are mostly related to the burning of multiple types of fuel.

- Stationary sources include smoke stacks of power plants, manufacturing facilities (factories) and waste incinerators, as well as furnaces and other types of fuel-burning heating devices. In developing and poor countries, traditional biomass burning is the major source of air pollutants; traditional biomass includes wood, crop waste and dung.
- Mobile sources include motor vehicles, marine vessels, and aircraft.
- Controlled burn practices in agriculture and forest management. Controlled or prescribed burning is a technique sometimes used in forest management, farming, prairie restoration or greenhouse gas abatement. Fire is a natural part of both forest and grassland ecology and controlled fire can be a tool for foresters. Controlled burning stimulates the germination of some desirable forest trees, thus renewing the forest.
- Fumes from paint, hair spray, varnish, aerosol sprays and other solvents
- Waste deposition in landfills, which generate methane. Methane is highly flammable and may form explosive mixtures with air. Methane is also an asphyxiant and may displace oxygen in an enclosed space. Asphyxia or suffocation may result if the oxygen concentration is reduced to below 19.5% by displacement.
- Military resources, such as nuclear weapons, toxic gases, germ warfare and rocketry

Natural sources:

Dust storm approaching Stratford, Texas.

- Dust from natural sources, usually large areas of land with little or no vegetation
- Methane, emitted by the digestion of food by animals, for example cattle
- Radon gas from radioactive decay within the Earth's crust. Radon is a colorless, odorless, naturally occurring, radioactive noble gas that is formed from the decay of radium. It is considered to be a health hazard. Radon gas from natural sources can accumulate in buildings, especially in confined areas such as the basement and it is the second most frequent cause of lung cancer, after cigarette smoking.
- Smoke and carbon monoxide from wildfires
- Vegetation, in some regions, emits environmentally significant amounts of Volatile organic compounds (VOCs) on warmer days. These VOCs react with primary anthropogenic pollutants—specifically, $NO_x$, $SO_2$, and anthropogenic organic carbon compounds — to produce a seasonal haze of secondary pollutants. Black gum, poplar, oak and willow are some examples of vegetation that can produce abundant VOCs. The VOC production from these species result in ozone levels up to eight times higher than the low-impact tree species.
- Volcanic activity, which produces sulfur, chlorine, and ash particulates

## Emission Factors

Beijing air on a 2005-day after rain (left) and a smoggy day (right)

Air pollutant emission factors are reported representative values that attempt to relate the quantity of a pollutant released to the ambient air with an activity associated with the release of that pollutant. These factors are usually expressed as the weight of pollutant divided by a unit weight, volume, distance, or duration of the activity emitting the pollutant (e.g., kilograms of particulate emitted per tonne of coal burned). Such factors facilitate estimation of emissions from various sources of air pollution. In most cases, these factors are simply averages of all available data of acceptable quality, and are generally assumed to be representative of long-term averages.

There are 12 compounds in the list of persistent organic pollutants. Dioxins and furans are two of them and intentionally created by combustion of organics, like open burning of plastics. These compounds are also endocrine disruptors and can mutate the human genes.

The United States Environmental Protection Agency has published a compilation of air pollutant emission factors for a wide range of industrial sources. The United Kingdom, Australia, Canada and many other countries have published similar compilations, as well as the European Environment Agency.

## Exposure

Air pollution risk is a function of the hazard of the pollutant and the exposure to that pollutant. Air pollution exposure can be expressed for an individual, for certain groups (e.g. neighborhoods or children living in a country), or for entire populations. For example, one may want to calculate the exposure to a hazardous air pollutant for a geographic area, which includes the various microenvironments and age groups. This can be calculated as an inhalation exposure. This would account for daily exposure in various settings (e.g. different indoor micro-environments and outdoor locations). The exposure needs to include different age and other demographic groups, especially infants, children, pregnant women and other sensitive subpopulations. The exposure to an air pollutant must integrate the concentrations of the air pollutant with respect to the time spent in each setting and the respective inhalation rates for each subgroup for each specific time that the subgroup is in the setting and engaged in particular activities (playing, cooking, reading, working, etc.). For example, a small child's inhalation rate will be less than that of an adult. A child engaged in vigorous exercise will have a higher respiration rate than the same child in a sedentary activity. The daily exposure, then, needs to reflect the time spent in each micro-environmental setting and the type of activities in these settings. The air pollutant concentration in each microactivity/micro-environmental setting is summed to indicate the exposure.

## Indoor Air Quality (IAQ)

Air quality monitoring, New Delhi, India.

A lack of ventilation indoors concentrates air pollution where people often spend the majority of their time. Radon (Rn) gas, a carcinogen, is exuded from the Earth in certain locations and trapped inside houses. Building materials including carpeting and plywood emit formaldehyde ($H_2CO$) gas. Paint and solvents give off volatile organic compounds (VOCs) as they dry. Lead paint can degenerate into dust and be inhaled. Intentional air pollution is introduced with the use of air fresheners, incense, and other scented items. Controlled wood fires in stoves and fireplaces can add significant amounts of smoke particulates into the air, inside and out. Indoor pollution fatalities may be caused by using pesticides and other chemical sprays indoors without proper ventilation.

Carbon monoxide poisoning and fatalities are often caused by faulty vents and chimneys, or by the burning of charcoal indoors or in a confined space, such as a tent. Chronic carbon monoxide poisoning can result even from poorly-adjusted pilot lights. Traps are built into all domestic plumbing to keep sewer gas and hydrogen sulfide, out of interiors. Clothing emits tetrachloroethylene, or other dry cleaning fluids, for days after dry cleaning.

Though its use has now been banned in many countries, the extensive use of asbestos in industrial and domestic environments in the past has left a potentially very dangerous material in many localities. Asbestosis is a chronic inflammatory medical condition affecting the tissue of the lungs. It occurs after long-term, heavy exposure to asbestos from asbestos-containing materials in structures. Sufferers have severe dyspnea (shortness of breath) and are at an increased risk regarding several different types of lung cancer. As clear explanations are not always stressed in non-technical literature, care should be taken to distinguish between several forms of relevant diseases. According to the World Health Organisation (WHO), these may defined as; asbestosis, *lung cancer*, and *Peritoneal Mesothelioma* (generally a very rare form of cancer, when more widespread it is almost always associated with prolonged exposure to asbestos).

Biological sources of air pollution are also found indoors, as gases and airborne particulates. Pets produce dander, people produce dust from minute skin flakes and decomposed hair, dust mites in bedding, carpeting and furniture produce enzymes and micrometre-sized fecal droppings, inhabitants emit methane, mold forms on walls and generates mycotoxins and spores, air conditioning systems can incubate Legionnaires' disease and mold, and houseplants, soil and surrounding gardens can produce pollen, dust, and mold. Indoors, the lack of air circulation allows these airborne pollutants to accumulate more than they would otherwise occur in nature.

## Health Effects

Air pollution is a significant risk factor for a number of pollution-related diseases and health conditions including respiratory infections, heart disease, COPD, stroke and lung cancer. The health effects caused by air pollution may include difficulty in breathing, wheezing, coughing, asthma and worsening of existing respiratory and cardiac conditions. These effects can result in increased medication use, increased doctor or emergency room visits, more hospital admissions and premature death. The human health effects of poor air quality are far reaching, but principally affect the body's respiratory system and the cardiovascular system. Individual reactions to air pollutants depend on the type of pollutant a person is exposed to, the degree of exposure, and the individual's health status and genetics. The most common sources of air pollution include particulates, ozone, nitrogen dioxide, and sulphur dioxide. Children aged less than five years that live in developing countries are the most vulnerable population in terms of total deaths attributable to indoor and outdoor air pollution.

## Mortality

The World Health Organization estimated in 2014 that every year air pollution causes the premature death of some 7 million people worldwide. India has the highest death rate due to air pollution. India also has more deaths from asthma than any other nation according to the World Health Organization. In December 2013 air pollution was estimated to kill 500,000 people in China each year. There is a positive correlation between pneumonia-related deaths and air pollution from motor vehicle emissions.

Annual premature European deaths caused by air pollution are estimated at 430,000. An important cause of these deaths is nitrogen dioxide and other nitrogen oxides (NOx) emitted by road vehicles. Across the European Union, air pollution is estimated to reduce life expectancy by almost nine months. Causes of deaths include strokes, heart disease, COPD, lung cancer, and lung infections.

Urban outdoor air pollution is estimated to cause 1.3 million deaths worldwide per year. Children are particularly at risk due to the immaturity of their respiratory organ systems.

The US EPA estimates that a proposed set of changes in diesel engine technology (*Tier 2*) could result in 12,000 fewer *premature mortalities*, 15,000 fewer heart attacks, 6,000 fewer emergency room visits by children with asthma, and 8,900 fewer respiratory-related hospital admissions each year in the United States.

The US EPA has estimated that limiting ground-level ozone concentration to 65 parts per billion, would avert 1,700 to 5,100 premature deaths nationwide in 2020 compared with the 75-ppb standard. The agency projected the more protective standard would also prevent an additional 26,000 cases of aggravated asthma, and more than a million cases of missed work or school. Following this assessment, the EPA acted to protect public health by lowering the National Ambient Air Quality Standards (NAAQS) for ground-level ozone to 70 parts per billion (ppb).

A new economic study of the health impacts and associated costs of air pollution in the Los Angeles Basin and San Joaquin Valley of Southern California shows that more than 3,800 people die prematurely (approximately 14 years earlier than normal) each year because air pollution levels violate federal standards. The number of annual premature deaths is considerably higher than the fatalities related to auto collisions in the same area, which average fewer than 2,000 per year.

Diesel exhaust (DE) is a major contributor to combustion-derived particulate matter air pollution. In several human experimental studies, using a well-validated exposure chamber setup, DE has been linked to acute vascular dysfunction and increased thrombus formation.

The mechanisms linking air pollution to increased cardiovascular mortality are uncertain, but probably include pulmonary and systemic inflammation.

## Cardiovascular Disease

A 2007 review of evidence found ambient air pollution exposure is a risk factor correlating with increased total mortality from cardiovascular events (range: 12% to 14% per 10 microg/m$^3$ increase).

Air pollution is also emerging as a risk factor for stroke, particularly in developing countries where pollutant levels are highest. A 2007 study found that in women, air pollution is not associated with hemorrhagic but with ischemic stroke. Air pollution was also found to be associated with increased incidence and mortality from coronary stroke in a cohort study in 2011. Associations are believed to be causal and effects may be mediated by vasoconstriction, low-grade inflammation and atherosclerosis Other mechanisms such as autonomic nervous system imbalance have also been suggested.

## Lung Disease

Chronic obstructive pulmonary disease (COPD) includes diseases such as chronic bronchitis and emphysema.

Research has demonstrated increased risk of developing asthma and COPD from increased exposure to traffic-related air pollution. Additionally, air pollution has been associated with increased hospitalization and mortality from asthma and COPD.

A study conducted in 1960-1961 in the wake of the Great Smog of 1952 compared 293 London residents with 477 residents of Gloucester, Peterborough, and Norwich, three towns with low reported death rates from chronic bronchitis. All subjects were male postal truck drivers aged 40 to 59. Compared to the subjects from the outlying towns, the London subjects exhibited more severe respiratory symptoms (including cough, phlegm, and dyspnea), reduced lung function ($FEV_1$ and peak flow rate), and increased sputum production and purulence. The differences were more pronounced for subjects aged 50 to 59. The study controlled for age and smoking habits, so concluded that air pollution was the most likely cause of the observed differences.

It is believed that much like cystic fibrosis, by living in a more urban environment serious health hazards become more apparent. Studies have shown that in urban areas patients suffer mucus hypersecretion, lower levels of lung function, and more self-diagnosis of chronic bronchitis and emphysema.

## Cancer

A review of evidence regarding whether ambient air pollution exposure is a risk factor for cancer in 2007 found solid data to conclude that long-term exposure to PM2.5 (fine particulates) increases the overall risk of non-accidental mortality by 6% per a 10 microg/$m^3$ increase. Exposure to PM2.5 was also associated with an increased risk of mortality from lung cancer (range: 15% to 21% per 10 microg/$m^3$ increase) and total cardiovascular mortality (range: 12% to 14% per a 10 microg/$m^3$ increase). The review further noted that living close to busy traffic appears to be associated with elevated risks of these three outcomes — increase in lung cancer deaths, cardiovascular deaths, and overall non-accidental deaths. The reviewers also found suggestive evidence that exposure to PM2.5 is positively associated with mortality from coronary heart diseases and exposure to $SO_2$ increases mortality from lung cancer, but the data was insufficient to provide solid conclusions. Another investigation showed that higher activity level increases deposition fraction of aerosol particles in human lung and recommended avoiding heavy activities like running in outdoor space at polluted areas.

In 2011, a large Danish epidemiological study found an increased risk of lung cancer for patients who lived in areas with high nitrogen oxide concentrations. In this study, the association was higher for non-smokers than smokers. An additional Danish study, also in 2011, likewise noted evidence of possible associations between air pollution and other forms of cancer, including cervical cancer and brain cancer.

In December 2015, medical scientists reported that cancer is overwhelmingly a result of environmental factors, and not largely down to bad luck. Maintaining a healthy weight, eating a healthy diet, minimizing alcohol and eliminating smoking reduces the risk of developing the disease, according to the researchers.

## Children

In the United States, despite the passage of the Clean Air Act in 1970, in 2002 at least 146 million Americans were living in non-attainment areas—regions in which the concentration of certain air pollutants exceeded federal standards. These dangerous pollutants are known as the criteria pollutants, and include ozone, particulate matter, sulfur dioxide, nitrogen dioxide, carbon monoxide, and lead. Protective measures to ensure children's health are being taken in cities such as New Delhi, India where buses now use compressed natural gas to help eliminate the "pea-soup" smog. A recent study in Europe has found that exposure to ultrafine particles can increase blood pressure in children.

## Infants

Ambient levels of air pollution have been associated with preterm birth and low birth weight. A 2014 WHO worldwide survey on maternal and perinatal health found a statistically significant association between low birth weights (LBW) and increased levels of exposure to PM2.5. Women in regions with greater than average PM2.5 levels had statistically significant higher odds of pregnancy resulting in a low-birth weight infant even when adjusted for country-related variables. The effect is thought to be from stimulating inflammation and increasing oxidative stress.

A study by the University of York found that in 2010 exposure to PM2.5 was strongly associated with 18% of preterm births globally, which was approximately 2.7 million premature births. The countries with the highest air pollution associated preterm births were in South and East Asia, the Middle East, North Africa, and West sub-Saharan Africa.

The source of PM 2.5 differs greatly by region. In South and East Asia, pregnant women are frequently exposed to indoor air pollution because of the wood and other biomass fuels used for cooking which are responsible for more than 80% of regional pollution. In the Middle East, North Africa and West sub-Saharan Africa, fine PM comes from natural sources, such as dust storms. The United States had an estimated 50,000 preterm births associated with exposure to PM2.5 in 2010.

A study performed by Wang, et al. between the years of 1988 and 1991 has found a correlation between Sulfur Dioxide ($SO_2$) and total suspended particulates (TSP) and preterm births and low birth weights in Beijing. A group of 74,671 pregnant women, in four separate regions of Beijing, were monitored from early pregnancy to delivery along with daily air pollution levels of Sulfur Dioxide and TSP (along with other particulates). The estimated reduction in birth weight was 7.3 g for every 100 μg/m3 increase in $SO_2$ and 6.9g for each 100 μg/m3 increase in TSP. These associations were statistically significant in both summer and winter, although, summer was greater. The proportion of low birth weight attributable to air pollution, was 13%. This is the largest attributable risk ever reported for the known risk factors of low birth weight. Coal stoves, which are in 97% of homes, are a major source of air pollution in this area.

Brauer et al. studied the relationship between air pollution and proximity to a highway with pregnancy outcomes in a Vancouver cohort of pregnant woman using addresses to estimate exposure during pregnancy. Exposure to NO, $NO_2$, CO $PM_{10}$ and PM2.5 were associated with infants born small for gestational age (SGA). Women living <50meters away from an expressway or highway were 26% more likely to give birth to a SGA infant.

## "Clean" Areas

Even in the areas with relatively low levels of air pollution, public health effects can be significant and costly, since a large number of people breathe in such pollutants. A 2005 scientific study for the British Columbia Lung Association showed that a small improvement in air quality (1% reduction of ambient PM2.5 and ozone concentrations) would produce $29 million in annual savings in the Metro Vancouver region in 2010. This finding is based on health valuation of lethal (death) and sub-lethal (illness) affects.

## Central Nervous System

Data is accumulating that air pollution exposure also affects the central nervous system.

In a June 2014 study conducted by researchers at the University of Rochester Medical Center, published in the journal Environmental Health Perspectives, it was discovered that early exposure to air pollution causes the same damaging changes in the brain as autism and schizophrenia. The study also shows that air pollution also affected short-term memory, learning ability, and impulsivity. Lead researcher Professor Deborah Cory-Slechta said that "When we looked closely at the ventricles, we could see that the white matter that normally surrounds them hadn't fully developed. It appears that inflammation had damaged those brain cells and prevented that region of the brain from developing, and the ventricles simply expanded to fill the space. Our findings add to the growing body of evidence that air pollution may play a role in autism, as well as in other neurodevelopmental disorders." Air pollution has a more significant negative effect on males than on females.

In 2015, experimental studies reported the detection of significant episodic (situational) cognitive impairment from impurities in indoor air breathed by test subjects who were not informed about changes in the air quality. Researchers at the Harvard University and SUNY Upstate Medical University and Syracuse University measured the cognitive performance of 24 participants in three different controlled laboratory atmospheres that simulated those found in "conventional" and "green" buildings, as well as green buildings with enhanced ventilation. Performance was evaluated objectively using the widely used Strategic Management Simulation software simulation tool, which is a well-validated assessment test for executive decision-making in an unconstrained situation allowing initiative and improvisation. Significant deficits were observed in the performance scores achieved in increasing concentrations of either volatile organic compounds (VOCs) or carbon dioxide, while keeping other factors constant. The highest impurity levels reached are not uncommon in some classroom or office environments.

## Agricultural Effects

In India in 2014, it was reported that air pollution by black carbon and ground level ozone had cut crop yields in the most affected areas by almost half in 2010 when compared to 1980 levels.

## Economic Effects

Air pollution costs the world economy $5 trillion per year as a result of productivity losses and degraded quality of life, according to a joint study by the World Bank and the Institute for Health Metrics and Evaluation (IHME) at the University of Washington These productivity losses are caused by deaths due to diseases caused by air pollution. One out of ten deaths in 2013 was caused by diseases associated with air pollution and the problem is getting worse. The problem is even more acute in the developing world. "Children under age 5 in lower-income countries are more than 60 times as likely to die from exposure to air pollution as children in high-income countries." The report states that additional economic losses caused by air pollution, including health costs and the adverse effect on agricultural and other productivity were not calculated in the report, and thus the actual costs to the world economy are far higher than $5 trillion.

## Historical Disasters

The world's worst short-term civilian pollution crisis was the 1984 Bhopal Disaster in India. Leaked industrial vapours from the Union Carbide factory, belonging to Union Carbide, Inc., U.S.A. (later bought by Dow Chemical Company), killed at least 3787 people and injured from 150,000 to 600,000. The United Kingdom suffered its worst air pollution event when the December 4 Great Smog of 1952 formed over London. In six days more than 4,000 died and more recent estimates put the figure at nearer 12,000. An accidental leak of anthrax spores from a biological warfare laboratory in the former USSR in 1979 near Sverdlovsk is believed to have caused at least 64 deaths. The worst single incident of air pollution to occur in the US occurred in Donora, Pennsylvania in late October, 1948, when 20 people died and over 7,000 were injured.

## Alternatives to Pollution

There are now practical alternatives to the principal causes of air pollution:

- Areas downwind (over 20 miles) of major airports more than double *total particulate emissions in air*, even when factoring in areas with frequent ship calls, and heavy freeway and city traffic like Los Angeles. Aviation biofuel mixed in with jetfuel at a 50/50 ratio can reduce jet derived cruise altitude particulate emissions by 50-70%, according to a NASA led 2017 study (however, this should imply ground level benefits to urban air pollution as well).
- Ship propulsion and idling can be switched to much cleaner fuels like natural gas. (Ideally a renewable source but not practical yet)
- Combustion of fossil fuels for space heating can be replaced by using ground source heat pumps and seasonal thermal energy storage.
- Electric power generation from burning fossil fuels can be replaced by power generation from nuclear and renewables. For poor nations, heating and home stoves that contribute much to regional air pollution can be replaced by a much cleaner fossil fuel like natural gas, or ideally, renewables.
- Motor vehicles driven by fossil fuels, a key factor in urban air pollution, can be replaced by electric vehicles. Though lithium supply and cost is a limitation, there are alternatives. Herding more people into clean public transit such as electric trains can also help. Nevertheless, even in emission-free electric vehicles, rubber tires produce significant amounts of air pollution themselves, ranking as 13th worst pollutant in Los Angeles.
- Biodigesters can be utilized in poor nations where slash and burn is prevalent, turning a useless commodity into a source of income. The plants can be gathered and sold to a central authority that will break it down in a large modern biodigester, producing much needed energy to use.

- Induced humidity and ventilation both can greatly dampen air pollution in enclosed spaces, which was found to be relatively high inside subway lines due to braking and friction and relatively less ironically inside transit buses than lower sitting passenger automobiles or subways.

## Reduction Efforts

There are various air pollution control technologies and strategies available to reduce air pollution. At its most basic level, land-use planning is likely to involve zoning and transport infrastructure planning. In most developed countries, land-use planning is an important part of social policy, ensuring that land is used efficiently for the benefit of the wider economy and population, as well as to protect the environment.

Because a large share of air pollution is caused by combustion of fossil fuels such as coal and oil, the reduction of these fuels can reduce air pollution drastically. Most effective is the switch to clean power sources such as wind power, solar power, hydro power which don't cause air pollution. Efforts to reduce pollution from mobile sources includes primary regulation (many developing countries have permissive regulations), expanding regulation to new sources (such as cruise and transport ships, farm equipment, and small gas-powered equipment such as string trimmers, chainsaws, and snowmobiles), increased fuel efficiency (such as through the use of hybrid vehicles), conversion to cleaner fuels or conversion to electric vehicles.

Titanium dioxide has been researched for its ability to reduce air pollution. Ultraviolet light will release free electrons from material, thereby creating free radicals, which break up VOCs and NOx gases. One form is superhydrophilic.

In 2014, Prof. Tony Ryan and Prof. Simon Armitage of University of Sheffield prepared a 10 meter by 20 meter-sized poster coated with microscopic, pollution-eating nanoparticles of titanium dioxide. Placed on a building, this giant poster can absorb the toxic emission from around 20 cars each day.

A very effective means to reduce air pollution is the transition to renewable energy. According to a study published in Energy and Environmental Science in 2015 the switch to 100% renewable energy in the United States would eliminate about 62,000 premature mortalities per year and about 42,000 in 2050, if no biomass were used. This would save about $600 billion in health costs a year due to reduced air pollution in 2050, or about 3.6% of the 2014 U.S. gross domestic product.

## Control Devices

The following items are commonly used as pollution control devices in industry and transportation. They can either destroy contaminants or remove them from an exhaust stream before it is emitted into the atmosphere.

- Particulate control
    - Mechanical collectors (dust cyclones, multicyclones)
    - Electrostatic precipitators An electrostatic precipitator (ESP), or electrostatic air cleaner is a particulate collection device that removes particles from a flowing gas (such as air), using the force of an induced electrostatic charge. Electrostatic pre-

cipitators are highly efficient filtration devices that minimally impede the flow of gases through the device, and can easily remove fine particulates such as dust and smoke from the air stream.

  - Baghouses Designed to handle heavy dust loads, a dust collector consists of a blower, dust filter, a filter-cleaning system, and a dust receptacle or dust removal system (distinguished from air cleaners which utilize disposable filters to remove the dust).
  - Particulate scrubbers Wet scrubber is a form of pollution control technology. The term describes a variety of devices that use pollutants from a furnace flue gas or from other gas streams. In a wet scrubber, the polluted gas stream is brought into contact with the scrubbing liquid, by spraying it with the liquid, by forcing it through a pool of liquid, or by some other contact method, so as to remove the pollutants.

- Scrubbers
  - Baffle spray scrubber
  - Cyclonic spray scrubber
  - Ejector venturi scrubber
  - Mechanically aided scrubber
  - Spray tower
  - Wet scrubber
- NOx control
  - Low NOx burners
  - Selective catalytic reduction (SCR)
  - Selective non-catalytic reduction (SNCR)
  - NOx scrubbers
  - Exhaust gas recirculation
  - Catalytic converter (also for VOC control)
- VOC abatement
  - Adsorption systems, using activated carbon, such as Fluidized Bed Concentrator
  - Flares
  - Thermal oxidizers
  - Catalytic converters
  - Biofilters
  - Absorption (scrubbing)

    - Cryogenic condensers
    - Vapor recovery systems
- Acid Gas/$SO_2$ control
    - Wet scrubbers
    - Dry scrubbers
    - Flue-gas desulfurization
- Mercury control
    - Sorbent Injection Technology
    - Electro-Catalytic Oxidation (ECO)
    - K-Fuel
- Dioxin and furan control
- Miscellaneous associated equipment
    - Source capturing systems
    - Continuous emissions monitoring systems (CEMS)

## Regulations

Smog in Cairo

In general, there are two types of air quality standards. The first class of standards (such as the U.S. National Ambient Air Quality Standards and E.U. Air Quality Directive) set maximum atmospheric concentrations for specific pollutants. Environmental agencies enact regulations which are intended to result in attainment of these target levels. The second class (such as the North American Air Quality Index) take the form of a scale with various thresholds, which is used to communicate to the public the relative risk of outdoor activity. The scale may or may not distinguish between different pollutants.

## Canada

In Canada, air pollution and associated health risks are measured with the Air Quality Health Index or (AQHI). It is a health protection tool used to make decisions to reduce short-term exposure to air pollution by adjusting activity levels during increased levels of air pollution.

The Air Quality Health Index or "AQHI" is a federal program jointly coordinated by Health Canada and Environment Canada. However, the AQHI program would not be possible without the commitment and support of the provinces, municipalities and NGOs. From air quality monitoring to health risk communication and community engagement, local partners are responsible for the vast majority of work related to AQHI implementation. The AQHI provides a number from 1 to 10+ to indicate the level of health risk associated with local air quality. Occasionally, when the amount of air pollution is abnormally high, the number may exceed 10. The AQHI provides a local air quality current value as well as a local air quality maximums forecast for today, tonight and tomorrow and provides associated health advice.

| | | | | | | | | | | |
|---|---|---|---|---|---|---|---|---|---|---|
| 1 | 2 | 3 | 4 | 5 | 6 | 7 | 8 | 9 | 10 | + |

Risk: Low (1-3) Moderate (4-6) High (7-10) Very high (above 10)

As it is now known that even low levels of air pollution can trigger discomfort for the sensitive population, the index has been developed as a continuum: The higher the number, the greater the health risk and need to take precautions. The index describes the level of health risk associated with this number as 'low', 'moderate', 'high' or 'very high', and suggests steps that can be taken to reduce exposure.

| **Health Risk** | **Air Quality Health Index** | **Health Messages** | |
|---|---|---|---|
| | | **At Risk population** | **General Population** |
| **Low** | **1-3** | **Enjoy** your usual outdoor activities. | **Ideal** air quality for outdoor activities |
| **Moderate** | **4-6** | **Consider reducing** or rescheduling strenuous activities outdoors if you are experiencing symptoms. | **No need to modify** your usual outdoor activities unless you experience symptoms such as coughing and throat irritation. |
| **High** | **7-10** | **Reduce** or reschedule strenuous activities outdoors. Children and the elderly should also take it easy. | **Consider reducing** or rescheduling strenuous activities outdoors if you experience symptoms such as coughing and throat irritation. |
| **Very high** | **Above 10** | **Avoid** strenuous activities outdoors. Children and the elderly should also avoid outdoor physical exertion and should stay indoors. | **Reduce** or reschedule strenuous activities outdoors, especially if you experience symptoms such as coughing and throat irritation. |

The measurement is based on the observed relationship of Nitrogen Dioxide ($NO_2$), ground-level Ozone ($O_3$) and particulates ($PM_{2.5}$) with mortality, from an analysis of several Canadian cities. Significantly, all three of these pollutants can pose health risks, even at low levels of exposure, especially among those with pre-existing health problems.

When developing the AQHI, Health Canada's original analysis of health effects included five major air pollutants: particulates, ozone, and nitrogen dioxide (NO2), as well as sulfur dioxide ($SO_2$), and carbon monoxide (CO). The latter two pollutants provided little information in predicting health effects and were removed from the AQHI formulation.

The AQHI does not measure the effects of odour, pollen, dust, heat or humidity.

## Germany

TA Luft is the German air quality regulation.

## Hotspots

Air pollution hotspots are areas where air pollution emissions expose individuals to increased negative health effects. They are particularly common in highly populated, urban areas, where there may be a combination of stationary sources (e.g. industrial facilities) and mobile sources (e.g. cars and trucks) of pollution. Emissions from these sources can cause respiratory disease, childhood asthma, cancer, and other health problems. Fine particulate matter such as diesel soot, which contributes to more than 3.2 million premature deaths around the world each year, is a significant problem. It is very small and can lodge itself within the lungs and enter the bloodstream. Diesel soot is concentrated in densely populated areas, and one in six people in the U.S. live near a diesel pollution hot spot.

While air pollution hotspots affect a variety of populations, some groups are more likely to be located in hotspots. Previous studies have shown disparities in exposure to pollution by race and/or income. Hazardous land uses (toxic storage and disposal facilities, manufacturing facilities, major roadways) tend to be located where property values and income levels are low. Low socioeconomic status can be a proxy for other kinds of social vulnerability, including race, a lack of ability to influence regulation and a lack of ability to move to neighborhoods with less environmental pollution. These communities bear a disproportionate burden of environmental pollution and are more likely to face health risks such as cancer or asthma.

Studies show that patterns in race and income disparities not only indicate a higher exposure to pollution but also higher risk of adverse health outcomes. Communities characterized by low socioeconomic status and racial minorities can be more vulnerable to cumulative adverse health impacts resulting from elevated exposure to pollutants than more privileged communities. Blacks and Latinos generally face more pollution than whites and Asians, and low-income communities bear a higher burden of risk than affluent ones. Racial discrepancies are particularly distinct in suburban areas of the US South and metropolitan areas of the US West. Residents in public housing, who are generally low-income and cannot move to healthier neighborhoods, are highly affected by nearby refineries and chemical plants.

## Cities

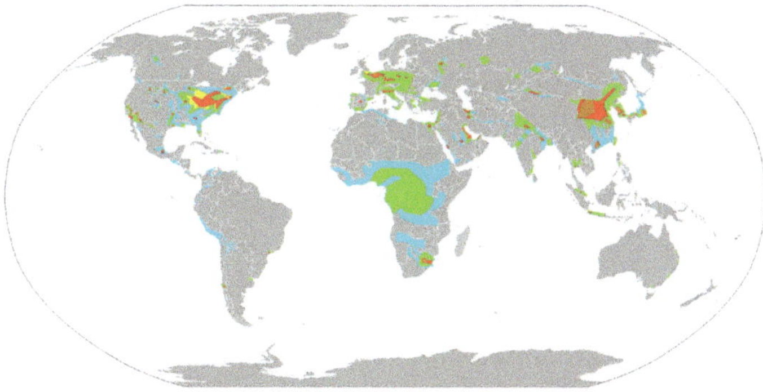

Nitrogen dioxide concentrations as measured from satellite 2002-2004

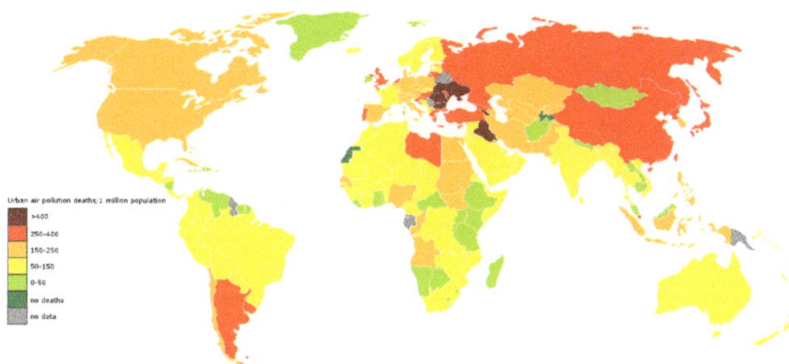


Deaths from air pollution in 2004

Air pollution is usually concentrated in densely populated metropolitan areas, especially in developing countries where environmental regulations are relatively lax or nonexistent. However, even populated areas in developed countries attain unhealthy levels of pollution, with Los Angeles and Rome being two examples. Between 2002 and 2011 the incidence of lung cancer in Beijing near doubled. While smoking remains the leading cause of lung cancer in China, the number of smokers is falling while lung cancer rates are rising. Another project focusing on the effects on pollution in vegetation has been researched by the local university in Sheffield, UK.

## National-scale air Toxics Assessments 1995–2005

The national-scale air toxics assessment (NATA) is an evaluation of air toxics by the U.S. EPA. The EPA has furnished four assessments that characterize nationwide chronic cancer risk estimates and noncancer hazards from inhaling air toxics. The lates was from 2005, and made publicly available in early 2011.

"EPA developed the NATA as a state-of-the-science screening tool for State/Local/Tribal Agencies to prioritize pollutants, emission sources and locations of interest for further study, in order to gain a better understanding of the risks. NATA assessments do not incorporate refined information about emission sources, but rather, use general information about sources to develop estimates of risks which are more likely to overestimate impacts than underestimate them. NATA provides estimates of the risk of cancer and other serious health effects from breathing (inhaling) air toxics in order to inform both national and more localized efforts to identify and prioritize air toxics, emission source types and locations which are of greatest potential concern in terms of contributing to population risk. This in turn helps air pollution experts focus limited analytical resources on areas and or populations where the potential for health risks are highest. Assessments include estimates of cancer and non-cancer health effects based on chronic exposure from outdoor sources, including assessments of non-cancer health effects for Diesel Particulate Matter. Assessments provide a snapshot of the outdoor air quality and the risks to human health that would result if air toxic emissions levels remained unchanged".

| **Most polluted cities by PM** | |
|---|---|
| **Particulate matter, μg/m³ (2004)** | **City** |
| 168 | Cairo, Egypt |
| 150 | Delhi, India |
| 128 | Kolkata, India (Calcutta) |
| 125 | Tianjin, China |
| 123 | Chongqing, China |
| 109 | Kanpur, India |
| 109 | Lucknow, India |
| 104 | Jakarta, Indonesia |
| 101 | Shenyang, China |

## Governing Urban Air Pollution

In Europe, Council Directive 96/62/EC on ambient air quality assessment and management provides a common strategy against which member states can "set objectives for ambient air quality in order to avoid, prevent or reduce harmful effects on human health and the environment . . . and improve air quality where it is unsatisfactory".

On 25 July 2008 in the case Dieter Janecek v Freistaat Bayern CURIA, the European Court of Justice ruled that under this directive citizens have the right to require national authorities to implement a short term action plan that aims to maintain or achieve compliance to air quality limit values.

This important case law appears to confirm the role of the EC as centralised regulator to European nation-states as regards air pollution control. It places a supranational legal obligation on the UK to protect its citizens from dangerous levels of air pollution, furthermore superseding national interests with those of the citizen.

In 2010, the European Commission (EC) threatened the UK with legal action against the successive breaching of PM10 limit values. The UK government has identified that if fines are imposed, they could cost the nation upwards of £300 million per year.

In March 2011, the Greater London Built-up Area remains the only UK region in breach of the EC's limit values, and has been given 3 months to implement an emergency action plan aimed at meeting the EU Air Quality Directive. The City of London has dangerous levels of PM10 concentrations, estimated to cause 3000 deaths per year within the city. As well as the threat of EU fines, in 2010 it was threatened with legal action for scrapping the western congestion charge zone, which is claimed to have led to an increase in air pollution levels.

In response to these charges, Boris Johnson, Mayor of London, has criticised the current need for European cities to communicate with Europe through their nation state's central government, arguing that in future "A great city like London" should be permitted to bypass its government and deal directly with the European Commission regarding its air quality action plan.

This can be interpreted as recognition that cities can transcend the traditional national government organisational hierarchy and develop solutions to air pollution using global governance networks, for example through transnational relations. Transnational relations include but are not exclusive to national governments and intergovernmental organisations, allowing sub-national actors including cities and regions to partake in air pollution control as independent actors.

Particularly promising at present are global city partnerships. These can be built into networks, for example the C40 Cities Climate Leadership Group, of which London is a member. The C40 is a public 'non-state' network of the world's leading cities that aims to curb their greenhouse emissions. The C40 has been identified as 'governance from the middle' and is an alternative to intergovernmental policy. It has the potential to improve urban air quality as participating cities "exchange information, learn from best practices and consequently mitigate carbon dioxide emissions independently from national government decisions". A criticism of the C40 network is that its exclusive nature limits influence to participating cities and risks drawing resources away from less powerful city and regional actors.

## Noise Pollution

Noise pollution or noise disturbance is the disturbing or excessive noise that may harm the activity or balance of human or animal life. The source of most outdoor noise worldwide is mainly caused by machines and transportation systems, motor vehicles engines, aircraft, and trains. Outdoor noise is summarized by the word environmental noise. Poor urban planning may give rise to noise pollution, since side-by-side industrial and residential buildings can result in noise pollution in the residential areas. Documented problems associated with urban noise go back as far as Ancient Rome.

Traffic is the main source of noise pollution in cities.

Outdoor noise can be caused by machines, construction activities, and music performances, especially in some workplaces. Noise-induced hearing loss can be caused by outside (e.g. trains) or inside (e.g. music) noise.

A Qantas Airways Boeing 747-400 passes close to houses shortly before landing at London Heathrow Airport.

High noise levels can contribute to cardiovascular effects in humans and an increased incidence of coronary artery disease. In animals, noise can increase the risk of death by altering predator or prey detection and avoidance, interfere with reproduction and navigation, and contribute to permanent hearing loss.

## Health

### Humans

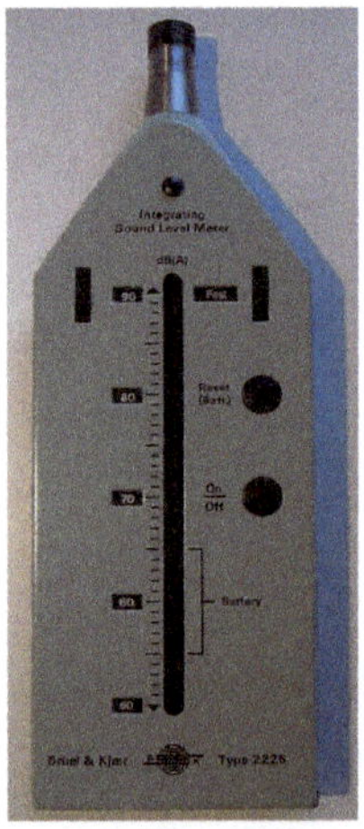

A sound level meter, a basic tool in measuring sound.

Noise pollution affects both health and behavior. Unwanted sound (noise) can damage psychological health. Noise pollution can cause hypertension, high stress levels, tinnitus, hearing loss, sleep disturbances, and other harmful effects.

Reaction to noise

Sound becomes unwanted when it either interferes with normal activities such as sleeping, conversation, or disrupts or diminishes one's quality of life.

Chronic exposure to noise may cause noise-induced hearing loss. Older males exposed to significant occupational noise demonstrate more significantly reduced hearing sensitivity than their non-exposed peers, though differences in hearing sensitivity decrease with time and the two groups are indistinguishable by age 79. A comparison of Maaban tribesmen, who were insignificantly exposed to transportation or industrial noise, to a typical U.S. population showed that chronic exposure to moderately high levels of environmental noise contributes to hearing loss.

High noise levels can result in cardiovascular effects and exposure to moderately high levels during a single eight-hour period causes a statistical rise in blood pressure of five to ten points and an increase in stress, and vasoconstriction leading to the increased blood pressure noted above, as well as to increased incidence of coronary artery disease.

Less addressed is how humans adapt to noise subjectively. Indeed, tolerance for noise is frequently independent of decibel levels. However, Murray Schafer's soundscape research was groundbreaking in this regard. In his eponymous work, he makes compelling arguments about how humans relate to noise on a subjective level, and how such subjectivity is conditioned by culture. He also notes that sound is an expression of power, and as such, material culture (e.g., fast cars or Harley Davidson motorcycles with aftermarket pipes) tend to have louder engines not only for safety reasons, but for expressions of power by dominating the soundscape with a particular sound. Other key research in this area can be seen in Fong's comparative analysis of soundscape differences between Bangkok, Thailand and Los Angeles, California, US. Fong's research methodology was modeled after Schafer, and the research findings show how not only do soundscapes differ, but they also rather explicitly point to the level of urban development in the area; that is, cities in the periphery - in Immanuel Wallerstein-speak - will have different soundscapes than that of cities in the core. Fong's important findings tie not only soundscape appreciation to our subjective views of sound, but also demonstrates how different sounds of the soundscape are indicative of class differences in urban environments.

## Wildlife

Noise can have a detrimental effect on wild animals, increasing the risk of death by changing the delicate balance in predator or prey detection and avoidance, and interfering the use of the sounds in communication, especially in relation to reproduction and in navigation. Acoustic overexposure can lead to temporary or permanent loss of hearing.

An impact of noise on wild animal life is the reduction of usable habitat that noisy areas may cause, which in the case of endangered species may be part of the path to extinction. Noise pollution may have caused the death of certain species of whales that beached themselves after being exposed to the loud sound of military sonar.

Noise also makes species communicate more loudly, which is called Lombard vocal response. Scientists and researchers have conducted experiments that show whales' song length is longer when submarine-detectors are on. If creatures do not "speak" loudly enough, their voice will be masked by anthropogenic sounds. These unheard voices might be warnings, finding of prey, or preparations of net-bubbling. When one species begins speaking more loudly, it will mask other species' voice, causing the whole ecosystem eventually to speak more loudly.

Marine invertebrates, such as crabs (*Carcinus maenas*), have also been shown to be negatively affected by ship noise. Larger crabs were noted to be negatively affected more by the sounds than smaller crabs. Repeated exposure to the sounds did lead to acclimatization.

European robins living in urban environments are more likely to sing at night in places with high levels of noise pollution during the day, suggesting that they sing at night because it is quieter, and their message can propagate through the environment more clearly. The same study showed that daytime noise was a stronger predictor of nocturnal singing than night-time light pollution, to which the phenomenon often is attributed.

Zebra finches become less faithful to their partners when exposed to traffic noise. This could alter a population's evolutionary trajectory by selecting traits, sapping resources normally devoted to other activities and thus leading to profound genetic and evolutionary consequences.

## Noise Mitigation

The sound tube in Melbourne, Australia is designed to reduce roadway noise without detracting from the area's aesthetics

Roadway noise can be reduced by the use of noise barriers, limitation of vehicle speeds, alteration of roadway surface texture, limitation of heavy vehicles, use of traffic controls that smooth vehicle flow to reduce braking and acceleration, and tire design. An important factor in applying these strategies is a computer model for roadway noise, that is capable of addressing local topography, meteorology, traffic operations, and hypothetical mitigation. Costs of building-in mitigation can be modest, provided these solutions are sought in the planning stage of a roadway project.

Aircraft noise can be reduced by using quieter jet engines. Altering flight paths and time of day runway has benefitted residents near airports.

Industrial noise has been addressed since the 1930s via redesign of industrial equipment, shock mounted assemblies and physical barriers in the workplace. In recent years, Buy Quiet programs and initiatives have arisen in an effort to combat occupational noise exposures. These programs promote the purchase of quieter tools and equipment and encourage manufacturers to design quieter equipment. The US National Institute for Occupational Health has created a database of industrial equipment with decibel levels noted.

## Legal Status

Up until the 1970s governments tended to view noise as a "nuisance" rather than an environmental problem.

Many conflicts over noise pollution are handled by negotiation between the emitter and the receiver. Escalation procedures vary by country, and may include action in conjunction with local authorities, in particular the police.

## India

Noise pollution is a major problem in India. The government of India has rules & regulations against firecrackers and loudspeakers, but enforcement is extremely lax. Awaaz Foundation is an Indian NGO working to control noise pollution from various sources through advocacy, public interest litigation, awareness, and educational campaigns since 2003. Despite increased enforcement and stringency of laws now being practised in urban areas, rural areas are still affected.

## United Kingdom

Figures compiled by rockwool, the mineral wool insulation manufacturer, based on responses from local authorities to a Freedom of Information Act (FOI) request reveal in the period April 2008 – 2009 UK councils received 315,838 complaints about noise pollution from private residences. This resulted in environmental health officers across the UK serving 8,069 noise abatement notices or citations under the terms of the Anti-Social Behaviour (Scotland) Act. In the last 12 months, 524 confiscations of equipment have been authorized involving the removal of powerful speakers, stereos and televisions. Westminster City Council has received more complaints per head of population than any other district in the UK with 9,814 grievances about noise, which equates to 42.32 complaints per thousand residents. Eight of the top 10 councils ranked by complaints per 1,000 residents are located in London.

## United States

There are federal standards for highway and aircraft noise; states and local governments typically have very specific statutes on building codes, urban planning, and roadway development.

Noise laws and ordinances vary widely among municipalities and indeed do not even exist in some cities. An ordinance may contain a general prohibition against making noise that is a nuisance, or it may set out specific guidelines for the level of noise allowable at certain times of the day and for certain activities.

The Environmental Protection Agency retains authority to investigate and study noise and its effect, disseminate information to the public regarding noise pollution and its adverse health effects, respond to inquiries on matters related to noise, and evaluate the effectiveness of existing regulations for protecting the public health and welfare, pursuant to the Noise Control Act of 1972 and the Quiet Communities Act of 1978.

New York City instituted the first comprehensive noise code in 1985. The Portland Noise Code includes potential fines of up to $5000 per infraction and is the basis for other major U.S. and Canadian city noise ordinances.

# Water Pollution

Raw sewage and industrial waste in the New River as it passes from Mexicali to Calexico, California

Water pollution is the contamination of water bodies (e.g. lakes, rivers, oceans, aquifers and groundwater). This form of environmental degradation occurs when pollutants are directly or indirectly discharged into water bodies without adequate treatment to remove harmful compounds.

Water pollution affects the entire biosphere – plants and organisms living in these bodies of water. In almost all cases the effect is damaging not only to individual species and population, but also to the natural biological communities.

## Introduction

Water pollution is a major global problem which requires ongoing evaluation and revision of water resource policy at all levels (international down to individual aquifers and wells). It has been suggested that water pollution is the leading worldwide cause of deaths and diseases, and that it accounts for the deaths of more than 14,000 people daily. An estimated 580 people in India die of water pollution related illness every day. About 90 percent of the water in the cities of China is polluted. As of 2007, half a billion Chinese had no access to safe drinking water. In addition to the acute problems of water pollution in developing countries, developed countries also continue to struggle with pollution problems. For example, in the most recent national report on water quality in the United States, 44 percent of assessed stream miles, 64 percent of assessed lake acres, and 30 percent of assessed bays and estuarine square miles were classified as polluted. The head of China's national development agency said in 2007 that one quarter the length of China's seven main rivers were so poisoned the water harmed the skin.

Water is typically referred to as polluted when it is impaired by anthropogenic contaminants and either does not support a human use, such as drinking water, or undergoes a marked shift in its ability to support its constituent biotic communities, such as fish. Natural phenomena such as volcanoes, algae blooms, storms, and earthquakes also cause major changes in water quality and the ecological status of water.

## Categories

Although interrelated, surface water and groundwater have often been studied and managed as separate resources. Surface water seeps through the soil and becomes groundwater. Conversely, groundwater can also feed surface water sources. Sources of surface water pollution are generally grouped into two categories based on their origin.

## Point Sources

Point source pollution at a shipyard in Rio de Janeiro, Brazil.

Point source water pollution refers to contaminants that enter a waterway from a single, identifiable source, such as a pipe or ditch. Examples of sources in this category include discharges from a sewage treatment plant, a factory, or a city storm drain. The U.S. Clean Water Act (CWA) defines point source for regulatory enforcement purposes. The CWA definition of point source was amended in 1987 to include municipal storm sewer systems, as well as industrial storm water, such as from construction sites.

## Non-point Sources

Nonpoint source pollution refers to diffuse contamination that does not originate from a single discrete source. NPS pollution is often the cumulative effect of small amounts of contaminants gathered from a large area. A common example is the leaching out of nitrogen compounds from fertilized agricultural lands. Nutrient runoff in storm water from "sheet flow" over an agricultural field or a forest are also cited as examples of NPS pollution.

Contaminated storm water washed off of parking lots, roads and highways, called urban runoff, is sometimes included under the category of NPS pollution. However, because this runoff is typically channeled into storm drain systems and discharged through pipes to local surface waters, it becomes a point source.

## Groundwater Pollution

Interactions between groundwater and surface water are complex. Consequently, groundwater pollution, also referred to as groundwater contamination, is not as easily classified as surface water pollution. By its very nature, groundwater aquifers are susceptible to contamination from sources that may not directly affect surface water bodies, and the distinction of point vs. non-point source may be irrelevant. A spill or ongoing release of chemical or radionuclide contaminants into soil (located away from a surface water body) may not create point or non-point source pollution but can contaminate the aquifer below, creating a toxic plume. The movement of the plume, called a plume front, may be analyzed through a hydrological transport model or groundwater model. Analysis of groundwater contamination may focus on soil characteristics and site geology, hydrogeology, hydrology, and the nature of the contaminants.

## Causes

The specific contaminants leading to pollution in water include a wide spectrum of chemicals, pathogens, and physical changes such as elevated temperature and discoloration. While many of the chemicals and substances that are regulated may be naturally occurring (calcium, sodium, iron, manganese, etc.) the concentration is often the key in determining what is a natural component of water and what is a contaminant. High concentrations of naturally occurring substances can have negative impacts on aquatic flora and fauna.

Oxygen-depleting substances may be natural materials such as plant matter (e.g. leaves and grass) as well as man-made chemicals. Other natural and anthropogenic substances may cause turbidity (cloudiness) which blocks light and disrupts plant growth, and clogs the gills of some fish species.

Many of the chemical substances are toxic. Pathogens can produce waterborne diseases in either human or animal hosts. Alteration of water's physical chemistry includes acidity (change in pH), electrical conductivity, temperature, and eutrophication. Eutrophication is an increase in the concentration of chemical nutrients in an ecosystem to an extent that increases the primary productivity of the ecosystem. Depending on the degree of eutrophication, subsequent negative environmental effects such as anoxia (oxygen depletion) and severe reductions in water quality may occur, affecting fish and other animal populations.

## Pathogens

A manhole cover unable to contain a sanitary sewer overflow.

Poster to teach people in South Asia about human activities leading to the pollution of water sources

Fecal sludge collected from pit latrines is dumped into a river at the Korogocho slum in Nairobi, Kenya.

Disease-causing microorganisms are referred to as pathogens. Although the vast majority of bacteria are either harmless or beneficial, a few pathogenic bacteria can cause disease. Coliform bacteria, which are not an actual cause of disease, are commonly used as a bacterial indicator of water pollution. Other microorganisms sometimes found in surface waters that have caused human health problems include:

- Burkholderia pseudomallei
- Cryptosporidium parvum
- Giardia lamblia
- Salmonella
- *Norovirus* and other viruses
- *Parasitic worms including the Schistosoma type*

High levels of pathogens may result from on-site sanitation systems (septic tanks, pit latrines) or inadequately treated sewage discharges. This can be caused by a sewage plant designed with less than secondary treatment (more typical in less-developed countries). In developed countries, older cities with aging infrastructure may have leaky sewage collection systems (pipes, pumps, valves), which can cause sanitary sewer overflows. Some cities also have combined sewers, which may discharge untreated sewage during rain storms.

Pathogen discharges may also be caused by poorly managed livestock operations.

## Organic, Inorganic and Macroscopic Contaminants

Contaminants may include organic and inorganic substances.

A garbage collection boom in an urban-area stream in Auckland, New Zealand.

Organic water pollutants include:

- Detergents
- Disinfection by-products found in chemically disinfected drinking water, such as chloroform
- Food processing waste, which can include oxygen-demanding substances, fats and grease
- Insecticides and herbicides, a huge range of organohalides and other chemical compounds
- Petroleum hydrocarbons, including fuels (gasoline, diesel fuel, jet fuels, and fuel oil) and lubricants (motor oil), and fuel combustion byproducts, from storm water runoff
- Volatile organic compounds, such as industrial solvents, from improper storage.
- Chlorinated solvents, which are dense non-aqueous phase liquids, may fall to the bottom of reservoirs, since they don't mix well with water and are denser.
    - Polychlorinated biphenyl (PCBs)
    - Trichloroethylene
- Perchlorate
- Various chemical compounds found in personal hygiene and cosmetic products

- Drug pollution involving pharmaceutical drugs and their metabolites

Inorganic water pollutants include:

- Acidity caused by industrial discharges (especially sulfur dioxide from power plants)
- Ammonia from food processing waste
- Chemical waste as industrial by-products
- Fertilizers containing nutrients--nitrates and phosphates—which are found in storm water runoff from agriculture, as well as commercial and residential use
- Heavy metals from motor vehicles (via urban storm water runoff) and acid mine drainage
- Secretion of creosote preservative into the aquatic ecosystem
- Silt (sediment) in runoff from construction sites, logging, slash and burn practices or land clearing sites.

Macroscopic pollution – large visible items polluting the water – may be termed "floatables" in an urban storm water context, or marine debris when found on the open seas, and can include such items as:

- Trash or garbage (e.g. paper, plastic, or food waste) discarded by people on the ground, along with accidental or intentional dumping of rubbish, that are washed by rainfall into storm drains and eventually discharged into surface waters
- Nurdles, small ubiquitous waterborne plastic pellets
- Shipwrecks, large derelict ships.

The Brayton Point Power Station in Massachusetts discharges heated water to Mount Hope Bay.

## Thermal Pollution

Thermal pollution is the rise or fall in the temperature of a natural body of water caused by human influence. Thermal pollution, unlike chemical pollution, results in a change in the physical properties of water. A common cause of thermal pollution is the use of water as a coolant by power plants and industrial manufacturers. Elevated water temperatures decrease oxygen levels, which can kill fish and alter food chain composition, reduce species biodiversity, and foster invasion by new thermophilic species. Urban runoff may also elevate temperature in surface waters.

Thermal pollution can also be caused by the release of very cold water from the base of reservoirs into warmer rivers.

## Transport and Chemical Reactions of Water Pollutants

Most water pollutants are eventually carried by rivers into the oceans. In some areas of the world the influence can be traced one hundred miles from the mouth by studies using hydrology transport models. Advanced computer models such as SWMM or the DSSAM Model have been used in many locations worldwide to examine the fate of pollutants in aquatic systems. Indicator filter-feeding species such as copepods have also been used to study pollutant fates in the New York Bight, for example. The highest toxin loads are not directly at the mouth of the Hudson River, but 100 km (62 mi) south, since several days are required for incorporation into planktonic tissue. The Hudson discharge flows south along the coast due to the coriolis force. Further south are areas of oxygen depletion caused by chemicals using up oxygen and by algae blooms, caused by excess nutrients from algal cell death and decomposition. Fish and shellfish kills have been reported, because toxins climb the food chain after small fish consume copepods, then large fish eat smaller fish, etc. Each successive step up the food chain causes a cumulative concentration of pollutants such as heavy metals (e.g. mercury) and persistent organic pollutants such as DDT. This is known as bio-magnification, which is occasionally used interchangeably with bio-accumulation.

A polluted river draining an abandoned copper mine on Anglesey

Large gyres (vortexes) in the oceans trap floating plastic debris. The North Pacific Gyre, for example, has collected the so-called "Great Pacific Garbage Patch", which is now estimated to be one hundred times the size of Texas. Plastic debris can absorb toxic chemicals from ocean pollution, potentially poisoning any creature that eats it. Many of these long-lasting pieces wind up in the stomachs of marine birds and animals. This results in obstruction of digestive pathways, which leads to reduced appetite or even starvation.

Many chemicals undergo reactive decay or chemical change, especially over long periods of time in groundwater reservoirs. A noteworthy class of such chemicals is the chlorinated hydrocarbons such as trichloroethylene (used in industrial metal degreasing and electronics manufacturing) and tetrachloroethylene used in the dry cleaning industry. Both of these chemicals, which are carcinogens themselves, undergo partial decomposition reactions, leading to new hazardous chemicals (including dichloroethylene and vinyl chloride).

Groundwater pollution is much more difficult to abate than surface pollution because groundwater can move great distances through unseen aquifers. Non-porous aquifers such as clays partially purify water of bacteria by simple filtration (adsorption and absorption), dilution, and, in some cases, chemical reactions and biological activity; however, in some cases, the pollutants merely transform to soil contaminants. Groundwater that moves through open fractures and caverns is not filtered and can be transported as easily as surface water. In fact, this can be aggravated by the human tendency to use natural sinkholes as dumps in areas of karst topography.

There are a variety of secondary effects stemming not from the original pollutant, but a derivative condition. An example is silt-bearing surface runoff, which can inhibit the penetration of sunlight through the water column, hampering photosynthesis in aquatic plants.

## Measurement

Environmental scientists preparing water autosamplers.

Water pollution may be analyzed through several broad categories of methods: physical, chemical and biological. Most involve collection of samples, followed by specialized analytical tests. Some methods may be conducted *in situ*, without sampling, such as temperature. Government agencies and research organizations have published standardized, validated analytical test methods to facilitate the comparability of results from disparate testing events.

## Sampling

Sampling of water for physical or chemical testing can be done by several methods, depending on the accuracy needed and the characteristics of the contaminant. Many contamination events are sharply restricted in time, most commonly in association with rain events. For this reason "grab" samples are often inadequate for fully quantifying contaminant levels. Scientists gathering this type of data often employ auto-sampler devices that pump increments of water at either time or discharge intervals.

Sampling for biological testing involves collection of plants and animals from the surface water body. Depending on the type of assessment, the organisms may be identified for biosurveys (population counts) and returned to the water body, or they may be dissected for bioassays to determine toxicity.

## Physical Testing

Common physical tests of water include temperature, solids concentrations (e.g., total suspended solids (TSS)) and turbidity.

## Chemical Testing

Water samples may be examined using the principles of analytical chemistry. Many published test methods are available for both organic and inorganic compounds. Frequently used methods include pH, biochemical oxygen demand (BOD), chemical oxygen demand (COD), nutrients (nitrate and phosphorus compounds), metals (including copper, zinc, cadmium, lead and mercury), oil and grease, total petroleum hydrocarbons (TPH), and pesticides.

## Biological Testing

Biological testing involves the use of plant, animal or microbial indicators to monitor the health of an aquatic ecosystem. They are any biological species or group of species whose function, population, or status can reveal what degree of ecosystem or environmental integrity is present. One example of a group of bio-indicators are the copepods and other small water crustaceans that are present in many water bodies. Such organisms can be monitored for changes (biochemical, physiological, or behavioral) that may indicate a problem within their ecosystem.

## Control of Pollution

Decisions on the type and degree of treatment and control of wastes, and the disposal and use of adequately treated wastewater, must be based on a consideration all the technical factors of each drainage basin, in order to prevent any further contamination or harm to the environment.

## Sewage Treatment

Deer Island Wastewater Treatment Plant serving Boston, Massachusetts and vicinity.

In urban areas of developed countries, domestic sewage is typically treated by centralized sewage treatment plants. Well-designed and operated systems (i.e., secondary treatment or better) can remove 90 percent or more of the pollutant load in sewage. Some plants have additional systems to remove nutrients and pathogens.

Cities with sanitary sewer overflows or combined sewer overflows employ one or more engineering approaches to reduce discharges of untreated sewage, including:

- utilizing a green infrastructure approach to improve storm water management capacity throughout the system, and reduce the hydraulic overloading of the treatment plant
- repair and replacement of leaking and malfunctioning equipment
- increasing overall hydraulic capacity of the sewage collection system (often a very expensive option).

A household or business not served by a municipal treatment plant may have an individual septic tank, which pre-treats the wastewater on site and infiltrates it into the soil.

## Industrial Wastewater Treatment

Dissolved air flotation system for treating industrial wastewater.

Some industrial facilities generate ordinary domestic sewage that can be treated by municipal facilities. Industries that generate wastewater with high concentrations of conventional pollutants (e.g. oil and grease), toxic pollutants (e.g. heavy metals, volatile organic compounds) or other non-conventional pollutants such as ammonia, need specialized treatment systems. Some of these facilities can install a pre-treatment system to remove the toxic components, and then send the partially treated wastewater to the municipal system. Industries generating large volumes of wastewater typically operate their own complete on-site treatment systems. Some industries have been successful at redesigning their manufacturing processes to reduce or eliminate pollutants, through a process called pollution prevention.

Heated water generated by power plants or manufacturing plants may be controlled with:

- cooling ponds, man-made bodies of water designed for cooling by evaporation, convection, and radiation
- cooling towers, which transfer waste heat to the atmosphere through evaporation or heat transfer

- cogeneration, a process where waste heat is recycled for domestic or industrial heating purposes.

Riparian buffer lining a creek in Iowa.

## Agricultural Wastewater Treatment

## Non Point Source Controls

Sediment (loose soil) washed off fields is the largest source of agricultural pollution in the United States. Farmers may utilize erosion controls to reduce runoff flows and retain soil on their fields. Common techniques include contour plowing, crop mulching, crop rotation, planting perennial crops and installing riparian buffers.

Feedlot in the United States

Nutrients (nitrogen and phosphorus) are typically applied to farmland as commercial fertilizer, animal manure, or spraying of municipal or industrial wastewater (effluent) or sludge. Nutrients may also enter runoff from crop residues, irrigation water, wildlife, and atmospheric deposition. Farmers can develop and implement nutrient management plans to reduce excess application of nutrients and reduce the potential for nutrient pollution.

To minimize pesticide impacts, farmers may use Integrated Pest Management (IPM) techniques (which can include biological pest control) to maintain control over pests, reduce reliance on chemical pesticides, and protect water quality.

## Point Source Wastewater Treatment

Farms with large livestock and poultry operations, such as factory farms, are called *concentrated animal feeding operations* or *feedlots* in the US and are being subject to increasing government regulation. Animal slurries are usually treated by containment in anaerobic lagoons before disposal by spray or trickle application to grassland. Constructed wetlands are sometimes used to facilitate treatment of animal wastes. Some animal slurries are treated by mixing with straw and composted at high temperature to produce a bacteriologically sterile and friable manure for soil improvement.

## Erosion and Sediment Control From Construction Sites

Sediment from construction sites is managed by installation of:

- erosion controls, such as mulching and hydroseeding, and
- sediment controls, such as sediment basins and silt fences.
- Discharge of toxic chemicals such as motor fuels and concrete washout is prevented by use of:
- spill prevention and control plans, and
- specially designed containers (e.g. for concrete washout) and structures such as overflow controls and diversion berms.

Silt fence installed on a construction site.

## Control of Urban Runoff (Storm Water)

Effective control of urban runoff involves reducing the velocity and flow of storm water, as well as reducing pollutant discharges. Local governments use a variety of storm water management techniques to reduce the effects of urban runoff. These techniques, called best management practices (BMPs) in the U.S., may focus on water quantity control, while others focus on improving water quality, and some perform both functions.

Retention basin for controlling urban runoff

Pollution prevention practices include low-impact development techniques, installation of green roofs and improved chemical handling (e.g. management of motor fuels & oil, fertilizers and pesticides). Runoff mitigation systems include infiltration basins, bioretention systems, constructed wetlands, retention basins and similar devices.

Thermal pollution from runoff can be controlled by storm water management facilities that absorb the runoff or direct it into groundwater, such as bioretention systems and infiltration basins. Retention basins tend to be less effective at reducing temperature, as the water may be heated by the sun before being discharged to a receiving stream.

## Thermal Pollution

The Brayton Point Power Station in Massachusetts discharges heated water to Mount Hope Bay.

Thermal pollution is the degradation of water quality by any process that changes ambient water temperature. A common cause of thermal pollution is the use of water as a coolant by power plants and industrial manufacturers. When water used as a coolant is returned to the natural environment at a higher temperature, the change in temperature decreases oxygen supply and affects ecosystem composition. Fish and other organisms adapted to particular temperature range can be killed by an abrupt change in water temperature (either a rapid increase or decrease) known as "thermal shock."

Urban runoff—stormwater discharged to surface waters from roads and parking lots—can also be a source of elevated water temperatures.

## Ecological Effects

Potrero Generating Station discharged heated water into San Francisco Bay. The plant was closed in 2011.

## Warm Water

Elevated temperature typically decreases the level of dissolved oxygen of water, as gases are less soluble in hotter liquids. This can harm aquatic animals such as fish, amphibians and other aquatic organisms. Thermal pollution may also increase the metabolic rate of aquatic animals, as enzyme activity, resulting in these organisms consuming more food in a shorter time than if their environment were not changed. An increased metabolic rate may result in fewer resources; the more adapted organisms moving in may have an advantage over organisms that are not used to the warmer temperature. As a result, food chains of the old and new environments may be compromised. Some fish species will avoid stream segments or coastal areas adjacent to a thermal discharge. Biodiversity can be decreased as a result.

High temperature limits oxygen dispersion into deeper waters, contributing to anaerobic conditions. This can lead to increased bacteria levels when there is ample food supply. Many aquatic species will fail to reproduce at elevated temperatures.

Primary producers (e.g. plants, cyanobacteria) are affected by warm water because higher water temperature increases plant growth rates, resulting in a shorter lifespan and species overpopulation. This can cause an algae bloom which reduces oxygen levels.

Temperature changes of even one to two degrees Celsius can cause significant changes in organism metabolism and other adverse cellular biology effects. Principal adverse changes can include rendering cell walls less permeable to necessary osmosis, coagulation of cell proteins, and alteration of enzyme metabolism. These cellular level effects can adversely affect mortality and reproduction.

A large increase in temperature can lead to the denaturing of life-supporting enzymes by breaking down hydrogen- and disulphide bonds within the quaternary structure of the enzymes. Decreased enzyme activity in aquatic organisms can cause problems such as the inability to break down lipids, which leads to malnutrition.

In limited cases, warm water has little deleterious effect and may even lead to improved function of the receiving aquatic ecosystem. This phenomenon is seen especially in seasonal waters and is known as thermal enrichment. An extreme case is derived from the aggregational habits of the manatee, which often uses power plant discharge sites during winter. Projections suggest that manatee populations would decline upon the removal of these discharges.

### Cold Water

Releases of unnaturally cold water from reservoirs can dramatically change the fish and macroinvertebrate fauna of rivers, and reduce river productivity. In Australia, where many rivers have warmer temperature regimes, native fish species have been eliminated, and macroinvertebrate fauna have been drastically altered. This may be mitigated by designing the dam to release warmer surface waters instead of the colder water at the bottom of the reservoir.

### Thermal Shock

When a power plant first opens or shuts down for repair or other causes, fish and other organisms adapted to particular temperature range can be killed by the abrupt change in water temperature, either an increase or decrease, known as "thermal shock."

## Sources and Control of Thermal Pollution

Cooling tower at Gustav Knepper Power Station, Dortmund, Germany

## Industrial Wastewater

In the United States, about 75 to 82 percent of thermal pollution is generated by power plants. The remainder is from industrial sources such as petroleum refineries, pulp and paper mills, chemical plants, steel mills and smelters. Heated water from these sources may be controlled with:

- cooling ponds, man-made bodies of water designed for cooling by evaporation, convection, and radiation
- cooling towers, which transfer waste heat to the atmosphere through evaporation and/or heat transfer
- cogeneration, a process where waste heat is recycled for domestic and/or industrial heating purposes.

Some facilities use once-through cooling (OTC) systems which do not reduce temperature as effectively as the above systems. For example, the Potrero Generating Station in San Francisco (closed in 2011), used OTC and discharged water to San Francisco Bay approximately 10 °C (20 °F) above the ambient bay temperature.

A bioretention cell for treating urban runoff in California

## Urban Runoff

During warm weather, urban runoff can have significant thermal impacts on small streams, as storm water passes over hot parking lots, roads and sidewalks. Storm water management facilities that absorb runoff or direct it into groundwater, such as bioretention systems and infiltration basins, can reduce these thermal effects. These and related systems for managing runoff are components of an expanding urban design approach commonly called green infrastructure.

Retention basins (stormwater ponds) tend to be less effective at reducing runoff temperature, as the water may be heated by the sun before being discharged to a receiving stream.

## Groundwater Pollution

Groundwater pollution (also called groundwater contamination) occurs when pollutants are released to the ground and make their way down into groundwater. It can also occur naturally due to the presence of a minor and unwanted constituent, contaminant or impurity in the groundwater, in which case it is more likely referred to as contamination rather than pollution.

Groundwater pollution example in Lusaka, Zambia where the pit latrine in the background is polluting the shallow well in the foreground with pathogens and nitrate.

The pollutant creates a contaminant plume within an aquifer. Movement of water and dispersion within the aquifer spreads the pollutant over a wider area. Its advancing boundary, often called a plume edge, can intersect with groundwater wells or daylight into surface water such as seeps and spring, making the water supplies unsafe for humans and wildlife. The movement of the plume, called a plume front, may be analyzed through a hydrological transport model or groundwater model. Analysis of groundwater pollution may focus on soil characteristics and site geology, hydrogeology, hydrology, and the nature of the contaminants.

Pollution can occur from on-site sanitation systems, landfills, effluent from wastewater treatment plants, leaking sewers, petrol filling stations or from over application of fertilizers in agriculture. Pollution (or contamination) can also occur from naturally occurring contaminants, such as arsenic or fluoride. Using polluted groundwater causes hazards to public health through poisoning or the spread of disease.

Different mechanisms have influence on the transport of pollutants, e.g. diffusion, adsorption, precipitation, decay, in the groundwater. The interaction of groundwater contamination with surface waters is analyzed by use of hydrology transport models.

## Pollutant Types

Contaminants found in groundwater cover a broad range of physical, inorganic chemical, organic chemical, bacteriological, and radioactive parameters. Principally, many of the same pollutants that play a role in surface water pollution may also be found in polluted groundwater, although their respective importance may differ.

## Arsenic and Fluoride

Arsenic and fluoride have been recognized by the World Health Organization (WHO) as the most serious inorganic contaminants in drinking-water on a worldwide basis.

The metalloid arsenic can occur naturally in groundwater, as seen most frequently in Asia, including in China, India and Bangladesh. In the Ganges Plain of northern India and Bangladesh severe contamination of groundwater by naturally occurring arsenic affects 25% of water wells in the shallower of two regional aquifers.

Arsenic in groundwater can also be present where there are mining operations or mine waste dumps that will leach arsenic.

Natural fluoride in groundwater is of growing concern as deeper groundwater is being used, "with more than 200 million people at risk of drinking water with elevated concentrations." Fluoride can especially be released from acidic volcanic rocks and dispersed volcanic ash when water hardness is low. High levels of fluoride in groundwater is a serious problem in the Argentinean Pampas, Chile, Mexico, India, Pakistan, the East African Rift, and some volcanic islands (Tenerife).

In areas that have naturally occurring high levels of fluoride in groundwater which is used for drinking water, both dental and skeletal fluorosis can be prevalent and severe.

## Pathogens

Waterborne diseases can be spread via a groundwater well which is contaminated with fecal pathogens from pit latrines

Pathogens contained in feces can lead to groundwater pollution when they are given the opportunity to reach the groundwater, making it unsafe for drinking. Of the four pathogen types that are present in feces (bacteria, viruses, protozoa and helminths or helminth eggs), the first three can be commonly found in polluted groundwater, whereas the relatively large helminth eggs are usually filtered out by the soil matrix.

Groundwater that is contaminated with pathogens can lead to fatal fecal-oral transmission of diseases (e.g. cholera, diarrhoea).

## Nitrate

Nitrate is the most common chemical contaminant in the world's groundwater and aquifers. In some low-income countries nitrate levels in groundwater are extremely high, causing significant health problems. It is also stable (it does not degrade) under high oxygen conditions.

Nitrate levels above 10 mg/L (10 ppm) in groundwater can cause "blue baby syndrome" (acquired methemoglobinemia). Drinking water quality standards in the European Union stipulate less than 50 mg/L for nitrate in drinking water.

However, the linkages between nitrates in drinking water and blue baby syndrome have been disputed in other studies. The syndrome outbreaks might be due to other factors than elevated nitrate concentrations in drinking water.

Elevated nitrate levels in groundwater can be caused by on-site sanitation, sewage sludge disposal and agricultural activities. It can therefore have an urban or agricultural origin.

## Organic Compounds

Volatile organic compounds (VOCs) are a dangerous contaminant of groundwater. They are generally introduced to the environment through careless industrial practices. Many of these compounds were not known to be harmful until the late 1960s and it was some time before regular testing of groundwater identified these substances in drinking water sources.

Primary VOC pollutants found in groundwater include aromatic hydrocarbons such as BTEX compounds ( benzene, toluene, ethylbenzene and xylenes), and chlorinated solvents including tetrachloroethylene (PCE), trichloroethylene (TCE), and vinyl chloride (VC). BTEX are important components of gasoline. PCE and TCE are industrial solvents historically used in dry cleaning processes and as a metal degreaser, respectively.

Other organic pollutants present in groundwater and derived from industrial operations are the polycyclic aromatic hydrocarbons (PAHs). Due to its molecular weight, Naphthalene is the most soluble and mobile PAH found in groundwater, whereas benzo(a)pyrene is the most toxic one. PAHs are generally produced as byproducts by incomplete combustion of organic matter.

Organic pollutants can also be found in groundwater as insecticides and herbicides. As many other synthetic organic compounds, most pesticides have very complex molecular structures. This complexity determines the water solubility, adsorption capacity, and mobility of pesticides in the groundwater system. Thus, some types of pesticides are more mobile than others so they can more easily reach a drinking-water source.

## Metals

Several trace metals can occurs naturally in some rocks or enter in the environmental from natural processes such as weathering. However, industrial activities such as mining and metallurgy, solid waste disposal, painting and enamel works can lead to severe groundwater pollution with elevated concentrations of toxic metals including lead, cadmium and chromium.

The migration of metals (and metalloids) in groundwater will be affected by several factors, in particular by chemical reactions which determine the partitioning of contaminants among different phases and species. Thus, the mobility of metals primarily depends on the pH and redox state of groundwater.

## Others

Other organic pollutants include a range of organohalides and other chemical compounds, petroleum hydrocarbons, various chemical compounds found in personal hygiene and cosmetic products, drug pollution involving pharmaceutical drugs and their metabolites. Inorganic pollutants might include other nutrients such as ammonia and phosphate, and radionuclides such as uranium (U) or radon (Rn) naturally present in some geological formations. Saltwater intrusion is also an example of natural contamination, but is very often intensified by human activities.

Groundwater pollution is a worldwide issue. A study of the groundwater quality of the principal aquifers of the United States conducted between 1991 and 2004, showed that 23% of domestic wells had contaminants at levels greater than human-health benchmarks. Another study suggested that the major groundwater pollution problems in Africa, considering the order of importance are: (1) nitrate pollution, (2) pathogenic agents, (3) organic pollution, (4) salinization, and (5) acid mine drainage.

## Causes

## Naturally-occurring (Geogenic)

"Geogenic" refers to naturally occurring as a result from geological processes.

The natural arsenic pollution occurs because aquifer sediments contain organic matter that generates anaerobic conditions in the aquifer. These conditions result in the microbial dissolution of iron oxides in the sediment and, thus, the release of the arsenic, normally strongly bound to iron oxides, into the water. As a consequence, arsenic-rich groundwater is often iron-rich, although secondary processes often obscure the association of dissolved arsenic and dissolved iron. Arsenic is found in groundwater most commonly as the reduced species arsenite and the oxidized species arsenate, being the acute toxicity of arsenite somewhat greater than that of arsenate. Investigations by WHO indicated that 20% of 25,000 boreholes tested in Bangladesh had arsenic concentrations exceeding 50 µg/l.

The occurrence of fluoride is close related to the abundance and solubility of fluoride-containing minerals such as fluorite (CaF2). Considerably high concentrations of fluoride in groundwater are typically caused by a lack of calcium in the aquifer. Health problems associated with dental fluorosis may occur when fluoride concentrations in groundwater exceed 1.5 mg/l, which is the WHO guideline value since 1984.

The Swiss Federal Institute of Aquatic Science and Technology (EAWAG) has recently developed the interactive Groundwater Assessment Platform (GAP), where the geogenic risk of contamination in a given area can be estimated using geological, topographical and other environmental data without having to test samples from every single groundwater resource. This tool also allows the user to produce probability risk mapping for both arsenic and fluoride.

High concentrations of parameters like salinity, iron, manganese, uranium, radon and chromium, in groundwater, may also be of geogenic origin. This contaminants can be important locally but they are not as widespread as arsenic and fluoride.

## On-site Sanitation Systems

Groundwater pollution with pathogens and nitrate can also occur from the liquids infiltrating into the ground from on-site sanitation systems such as pit latrines and septic tanks, depending on the population density and the hydrogeological conditions.

A traditional housing compound near Herat, Afghanistan, where a shallow water supply well (foreground) is in close proximity to the pit latrine (behind the white greenhouse) leading to contamination of the groundwater.

Factors controlling the fate and transport of pathogens are quite complex and the interaction among them is not well understood. If the local hydrogeological conditions (which can vary within a space of a few square kilometres) are ignored, simple on-site sanitation infrastructures such as pit latrines can cause significant public health risks via contaminated groundwater.

Liquids leach from the pit and pass the unsaturated soil zone (which is not completely filled with water). Subsequently, these liquids from the pit enter the groundwater where they may lead to groundwater pollution. This is a problem if a nearby water well is used to supply groundwater for drinking water purposes. During the passage in the soil, pathogens can die off or be adsorbed significantly, mostly depending on the travel time between the pit and the well. Most, but not all pathogens die within 50 days of travel through the subsurface.

The degree of pathogen removal strongly varies with soil type, aquifer type, distance and other environmental factors. For example, the unsaturated zone becomes "washed" during extended periods of heavy rain, providing hydraulic pathway for the quick pass of pathogens. It is difficult to estimate the safe distance between a pit latrine or a septic tank and a water source. In any case, such recommendations about the safe distance are mostly ignored by those building pit latrines. In addition, household plots are of a limited size and therefore pit latrines are often built much closer to groundwater wells than what can be regarded as safe. This results in groundwater pollution and household members falling sick when using this groundwater as a source of drinking water.

## Sewage (Treated and Untreated)

Groundwater pollution can be caused by untreated waste discharge leading to diseases like skin lesions, bloody diarrhea and dermatitis. This is more common in locations having limited wastewater treatment infrastructure, or where there are systematic failures of the on-site sewage disposal system. Along with pathogens and nutrients, untreated sewage can also have an important load of heavy metals that may seep into the groundwater system.

The treated effluent from sewage treatment plants may also reach the aquifer if the effluent is infiltrated or discharged to local surface water bodies. Therefore, those substances that are not removed in conventional sewage treatment plants may reach the groundwater as well. For example, detected concentrations of pharmaceutical residues in groundwater were in the order of 50 ng/L in several locations in Germany. This is because in conventional sewage treatment plants, micro-pollutants such as hormones, pharmaceutical residues and other micro-pollutants contained in urine and feces are only partially removed and the remainder is discharged into surface water, from where it may also reach the groundwater.

Groundwater pollution can also occur from leaking sewers which has been observed for example in Germany. This can also lead to potential cross-contamination of drinking-water supplies.

Spreading wastewater or sewage sludge in agriculture may also be included as sources of faecal contamination in groundwater.

## Fertilizers and Pesticides

Nitrate can also enter the groundwater via excessive use of fertilizers, including manure spreading. This is because only a fraction of the nitrogen-based fertilizers is converted to produce and other plant matter. The remainder accumulates in the soil or lost as run-off. High application rates of nitrogen-containing fertilizers combined with the high water-solubility of nitrate leads to increased runoff into surface water as well as leaching into groundwater, thereby causing groundwater pollution. The excessive use of nitrogen-containing fertilizers (be they synthetic or natural) is particularly damaging, as much of the nitrogen that is not taken up by plants is transformed into nitrate which is easily leached.

Poor management practices in manure spreading can introduce both pathogens and nutrients (nitrate) in the groundwater system.

The nutrients, especially nitrates, in fertilizers can cause problems for natural habitats and for human health if they are washed off soil into watercourses or leached through soil into groundwater.

The heavy use of nitrogenous fertilizers in cropping systems is the largest contributor to anthropogenic nitrogen in groundwater worldwide.

Feedlots/animal corrals can also lead to the potential leaching of nitrogen and metals to groundwater. Over application of animal manure may also result in groundwater pollution with pharmaceutical residues derived from veterinary drugs.

The US Environmental Protection Agency (EPA) and the European Commission are seriously dealing with the nitrate problem related to agricultural development, as a major water supply problem that requires appropriate management and governance.

Runoff of pesticides may leach into groundwater causing human health problems from contaminated water wells. Pesticide concentrations found in groundwater are typically low, and often the regulatory human health-based limits exceeded are also very low. The organophosphorus insecticide monocrotophos (MCP) appears to be one of a few hazardous, persistent, soluble and mobile (it does not bind with minerals in soils) pesticides able to reach a drinking-water source. In general, more pesticide compounds are being detected as groundwater quality monitoring programs have become more extensive; however, much less monitoring has been conducted in developing countries due to the high analysis costs.

## Commercial and Industrial Leaks

A wide variety of both inorganic and organic pollutants have been found in aquifers underlying commercial and industrial activities.

Ore mining and metal processing facilities are the primary responsible of the presence of metals in groundwater of anthropogenic origin, including arsenic. The low pH associated with acid mine drainage (AMD) contributes to the solubility of potential toxic metals that can eventually enter the groundwater system.

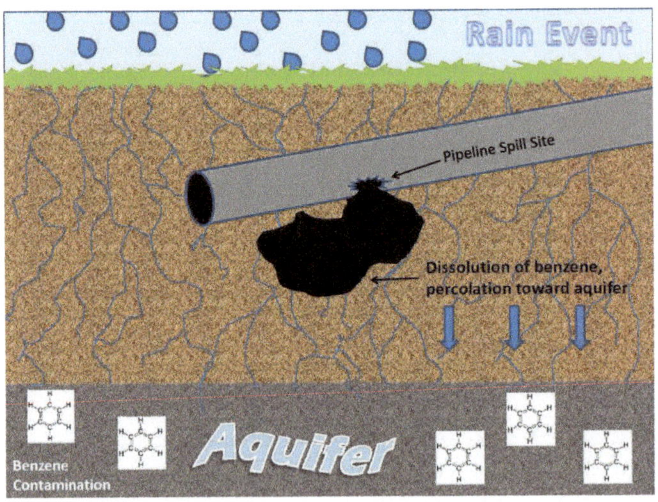


Oil spills associa ted with underground pipelines and tanks can release benzene and other soluble petroleum hydrocarbons that rapidly percolate down into the aquifer.

There is an increasing concern over the groundwater pollution by gasoline leaked from petroleum underground storage tanks (USTs) of gas stations. BTEX compounds are the most common additives of the gasoline. BTEX compounds, including benzene, have densities lower than water (1 g/ml). Similar to the oil spills on the sea, the non-miscible phase, referred to as Light Non-Aqueous Phase Liquid (LNAPL), will "float" upon the water table in the aquifer.

Chlorinated solvents are used in nearly any industrial practice where degreasing removers are required. PCE is a highly utilized solvent in the dry cleaning industry because of its cleaning effectiveness and relatively low cost. It has also been used for metal-degreasing operations. Because it is highly volatile, it is more frequently found in groundwater than in surface water. TCE has historically been used as a metal cleaning. The military facility Anniston Army Dept (ANAD) in the United States was placed on the US EPA Superfund National Priorities List (NPL) because of groundwater contamination with as much as 27 million pounds of TCE. Both PCE and TCE may degrade to vinyl chloride (VC), the most toxic chlorinated hydrocarbon.

Many types of solvents may have also been disposed illegally, leaking over time to the groundwater system.

Chlorinated solvents such as PCE and TCE have densities higher than water and the non-miscible phase is referred to as Dense Non-Aqueous Phase Liquids (DNAPL). Once they reach the aquifer, they will "sink" and eventually accumulate on the top of low-permeability layers.

Historically, wood-treating facilities have also release insecticides such as pentachlorophenol (PCP) and creosote into the environment, impacting the groundwater resources. PCP is a highly soluble and toxic obsolete pesticide recently listed in the Stockholm Convention on Persistent Organic Pollutants. PAHs and other semi-VOCs are the common contaminants associated with creosote.

Although non-miscible, both LNAPLs and DNAPLs still have the potential to slowly dissolve into the aqueous (miscible) phase to create a plume and thus become a long-term source of contamination. DNAPLs (chlorinated solvents, heavy PAHs, creosote, PCBs) tend to be difficult to manage as they can reside very deep in the groundwater system.

## Hydraulic Fracturing

The recent growth of Hydraulic Fracturing ("Fracking") wells in the United States has raised concerns regarding its potential risks of contaminating groundwater resources. The Environmental Protection Agency (EPA), along with many other researchers, has been delegated to study the relationship between hydraulic fracturing and drinking water resources. While it is possible to perform hydraulic fracturing without having a relevant impact on groundwater resources if stringent controls und quality management measures are in place, there are a number of cases where groundwater pollution due to improper handling or technical failures was observed.

While the EPA has not found significant evidence of a widespread, systematic impact on drinking water by hydraulic fracturing, this may be due to insufficient systematic pre- and post- hydraulic fracturing data on drinking water quality, and the presence of other agents of contamination that preclude the link between shale oil/gas extraction and its impact.

Despite the EPA's lack of profound widespread evidence, other researchers have made significant observations of rising groundwater contamination in close proximity to major shale oil/

gas drilling sites located in Marcellus*Ellsworth, William (2013). "Injection-Induced Earthquakes". Science AAAS.*</ref> (British Columbia, Canada). Within one kilometer of these specific sites, a subset of shallow drinking water consistently showed higher concentration levels of methane, ethane, and propane concentrations than normal. An evaluation of higher Helium and other noble gas concentration along with the rise of hydrocarbon levels supports the distinction between hydraulic fracturing fugitive gas and naturally occurring "background" hydrocarbon content. This contamination is speculated to be the result of leaky, failing, or improperly installed gas well casings.

Furthermore, it is theorized that contamination could also result from the capillary migration of deep residual hyper-saline water and hydraulic fracturing fluid, slowly flowing through faults and fractures until finally making contact with groundwater resources; however, many researchers argue that the permeability of rocks overlying shale formations are too low to allow this to ever happen sufficiently. To ultimately prove this theory, there would have to be traces of toxic trihalomethanes (THM) since they are often associated with the presence of stray gas contamination, and typically co-occur with high halogen concentrations in hyper-saline waters. Besides, highly saline waters are a common natural feature in deep groundwater systems.

While conclusions regarding groundwater pollution as the result to hydraulic fracturing fluid flow is restricted in both space and time, researchers have hypothesized that the potential for systematic stray gas contamination depends mainly on the integrity of the shale oil/gas well structure, along with its relative geological location to local fracture systems that could potentially provide flow paths for fugitive gas migration.

Though widespread, systematic contamination by hydraulic fracturing has been heavily disputed, one major source of contamination that has the most consensus among researchers of being the most problematic is site-specific accidental spillage of hydraulic fracturing fluid and produced water. So far, a significant majority of groundwater contamination events are derived from surface-level anthropogenic routes rather than the subsurface flow from underlying shale formations. Examples of such events include: a fracking fluid spillage in Acorn Fork Creek, Kentucky that caused a widespread death among aquatic species in 2007; a 420,000 gallon spillage of hyper-saline produced water that turned a once very-fertile farmland in New Mexico into a dead-zone in 2010; and a 42,000 gallon fracking fluid spillage in Arlington, Texas that necessitated an evacuation of over a 100 homes in 2015. While the damage can be obvious, and much more effort is being done to prevent these accidents from occurring so frequently, the lack of data from fracking oil spills continue to leave researchers in the dark. In many of these events, the data acquired from the leakage or spillage is often very vague, and thus would lead researchers to lacking conclusions.

Researchers from the Federal Institute for Geosciences and Natural Resources (BGR) conducted a modelling study for a deep shale-gas formation in the North German Basin. They concluded that the probability is small that the rise of fracking fluids through the geological underground to the surface will impact shallow groundwater.

## Landfill Leachate

Leachate from sanitary landfills can lead to groundwater pollution.

Love Canal was one of the most widely known examples of groundwater pollution. In 1978, residents of the Love Canal neighborhood in upstate New York noticed high rates of cancer and an alarming number of birth defects. This was eventually traced to organic solvents and dioxins from an industrial landfill that the neighborhood had been built over and around, which had then infiltrated into the water supply and evaporated in basements to further contaminate the air. Eight hundred families were reimbursed for their homes and moved, after extensive legal battles and media coverage.

## Other

Further causes of groundwater pollution are chemical spills from commercial or industrial operations, chemical spills occurring during transport (e.g. spillage of diesel fuels), illegal waste dumping, infiltration from urban runoff or mining operations, road salts, de-icing chemicals from airports and even atmospheric contaminants since groundwater is part of the hydrologic cycle.

The burial of corpses and their subsequent degradation may also pose a risk of pollution to groundwater.

## Mechanisms

The passage of water through the subsurface can provide a reliable natural barrier to contamination but it only works under favorable conditions.

The stratigraphy of the area plays an important role in the transport of pollutants. An area can have layers of sandy soil, fractured bedrock, clay, or hardpan. Areas of karst topography on limestone bedrock are sometimes vulnerable to surface pollution from groundwater. Earthquake faults can also be entry routes for downward contaminant entry. Water table conditions are of great importance for drinking water supplies, agricultural irrigation, waste disposal (including nuclear waste), wildlife habitat, and other ecological issues.

## Interactions with Surface Water

Although interrelated, surface water and groundwater have often been studied and managed as separate resources. Surface water seeps through the soil and becomes groundwater. Conversely, groundwater can also feed surface water sources. Sources of surface water pollution are generally grouped into two categories based on their origin.

Interactions between groundwater and surface water are complex. Consequently, groundwater pollution, sometimes referred to as groundwater contamination, is not as easily classified as surface water pollution. By its very nature, groundwater aquifers are susceptible to contamination from sources that may not directly affect surface water bodies, and the distinction of point vs. non-point source may be irrelevant. A spill or ongoing release of chemical or radionuclide contaminants into soil (located away from a surface water body) may not create point or non-point source pollution but can contaminate the aquifer below, creating a toxic plume.

## Prevention

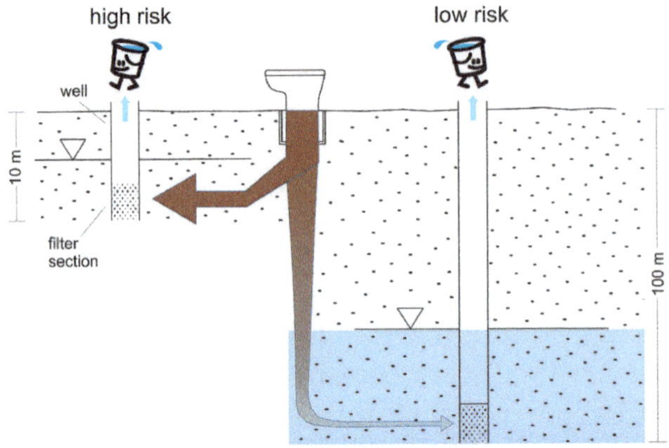


Schematic showing that there is a lower risk of groundwater pollution with greater depth of the water well

## Precautionary Principle

The precautionary principle, evolved from Principle 15 of the Rio Declaration on Environment and Development, is important in protecting groundwater resources from pollution. The precautionary principle provides that "*where there are threats of irreversible damage, lack of full scientific certainty shall not be used as reason for postponing cost-effective measures to prevent environmental degradation*".

One of the six basic principles of the European Union (EU) water policy is the application of the precautionary principle.

## Groundwater Quality Monitoring

Groundwater quality monitoring programs have been implemented regularly in many countries around the world. They are important components to understand the hydrogeological system, and for the development of conceptual models and aquifer vulnerability maps.

Groundwater quality must be regularly monitored across the aquifer to determine trends. Effective groundwater monitoring should be driven by a specific objective, for example, a specific contaminant of concern. Contaminant levels can be compared to the World Health Organization (WHO) guidelines for drinking-water quality. It is not rare that limits of contaminants are reduced as more medical experience is gained.

Sufficient investment should be given to continue monitoring over the long term. When a problem is found, action should be taken to correct it. Waterborne outbreaks in the United States decreased with the introduction of more stringent monitoring (and treatment) requirements in the early 90s.

The community can also help monitor the groundwater quality.

## Land Zoning for Ground water Protection

The development of land-use zoning maps has been implemented by several water authorities at different scales around the world. There are two types of zoning maps: aquifer vulnerability maps and source protection maps.

## Aquifer Vulnerability Map

It refers to the intrinsic (or natural) vulnerability of a groundwater system to pollution. Intrinsically, some aquifers are more vulnerable to pollution than other aquifers. Shallow unconfined aquifers are more at risk of pollution because there are fewer layers to filter out contaminants.

The unsaturated zone can play an important role in retarding (and in some cases eliminating) pathogens and so must be considered when assessing aquifer vulnerability. The biological activity is greatest in the top soil layers where the attenuation of pathogens is generally most effective.

Preparation of the vulnerability maps typically involves overlaying several thematic maps of physical factors that have been selected to describe the aquifer vulnerability. The index-based parametric mapping method GOD developed by Foster and Hirata (1988) uses three generally available or readily estimated parameters, the degree of Groundwater hydraulic confinement, geological nature of the Overlying strata and Depth to groundwater. A further approach developed by the US EPA named DRASTIC employs seven hydrogeological factors to develop an index of vulnerability: Depth to water table, net Recharge, Aquifer media, Soil media, Topography (slope), Impact on the vadose zone, and hydraulic Conductivity.

There is a particular debate among hydrogeologist whether aquifer vulnerability should be established in a general (intrinsic) way for all contaminants, or specifically for each pollutant.

## Source Protection Map

It refers to the capture areas around an individual groundwater source, such as a water well or a spring, to especially protect them from pollution. Thus, potential sources of degradable pollutants, such as pathogens, can be located at distances which travel times along the flowpaths are long enough for the pollutant to be eliminated through filtration or adsorption.

Analytical methods using equations to define groundwater flow and contaminant transport are the most widely used. The WHPA is a semi-analytical groundwater flow simulation program developed by the US EPA for delineating capture zones in a wellhead protection area.

The simplest form of zoning employs fixed-distance methods where activities are excluded within a uniformly applied specified distance around abstraction points.

## Locating on-site Sanitation Systems

As the health effects of most toxic chemicals arise after prolonged exposure, risk to health from chemicals is generally lower than that from pathogens. Thus, the quality of the source protection measures is an important component in controlling whether pathogens may be present in the final drinking-water.

On-site sanitation systems can be designed in such a way that groundwater pollution from these sanitation systems is prevented from occurring. Detailed guidelines have been developed to estimate safe distances to protect groundwater sources from pollution from on-site sanitation. The following criteria have been proposed for safe siting (i.e. deciding on the location) of on-site sanitation systems:

- Horizontal distance between the drinking water source and the sanitation system
    - Guideline values for horizontal separation distances between on-site sanitation systems and water sources vary widely (e.g. 15 to 100 m horizontal distance between pit latrine and groundwater wells)
- Vertical distance between drinking water well and sanitation system
- Aquifer type
- Groundwater flow direction
- Impermeable layers
- Slope and surface drainage
- Volume of leaking wastewater
- Superposition, i.e. the need to consider a larger planning area

As a very general guideline it is recommended that the bottom of the pit should be at least 2 m above groundwater level, and a minimum horizontal distance of 30 m between a pit and a water source is normally recommended to limit exposure to microbial contamination. However, no general statement should be made regarding the minimum lateral separation distances required to prevent contamination of a well from a pit latrine. For example, even 50 m lateral separation distance might not be sufficient in a strongly karstified system with a downgradient supply well or spring, while 10 m lateral separation distance is completely sufficient if there is a well developed clay cover layer and the annular space of the groundwater well is well sealed.

## Legislation

Institutional and legal issues are critical in determining the success or failure of groundwater protection policies and strategies.

Sign near Mannheim, Germany indicating a zone as a dedicated "groundwater protection zone"

## United States

In November 2006, the Environmental Protection Agency published the Ground Water Rule in the United States Federal Register. The EPA was worried that the ground water system would be vulnerable to contamination from fecal matter. The point of the rule was to keep microbial pathogens out of public water sources. The 2006 Ground Water Rule was an amendment of the 1996 Safe Drinking Water Act.

The ways to deal with groundwater pollution that has already occurred can be grouped into the following categories: containing the pollutants to prevent them from migrating further; removing the pollutants from the aquifer; remediating the aquifer by either immobilizing or detoxifying the contaminants while they are still in the aquifer (in-situ); treating the groundwater at its point of use; or abandoning the use of this aquifer's groundwater and finding an alternative source of water.

## Management

### Point-of-use Treatment

Portable water purification devices or "point-of-use" (POU) water treatment systems and field water disinfection techniques can be used to remove some forms of groundwater pollution prior to drinking, namely any fecal pollution. Many commercial portable water purification systems or chemical additives are available which can remove pathogens, chlorine, bad taste, odors, and heavy metals like lead and mercury.

Techniques include boiling, filtration, activated charcoal absorption, chemical disinfection, ultraviolet purification, ozone water disinfection, solar water disinfection, solar distillation, homemade water filters.

Arsenic removal filters (ARF) are dedicated technologies typically installed to remove arsenic. Many of these technologies require a capital investment and long-term maintenance. Filters in Bangladesh are usually abandoned by the users due to their high cost and complicated maintenance, which is also quite expensive.

### Groundwater Remediation

Groundwater pollution is much more difficult to abate than surface pollution because groundwater can move great distances through unseen aquifers. Non-porous aquifers such as clays partially purify water of bacteria by simple filtration (adsorption and absorption), dilution, and, in some cases, chemical reactions and biological activity; however, in some cases, the pollutants merely transform to soil contaminants. Groundwater that moves through open fractures and caverns is not filtered and can be transported as easily as surface water. In fact, this can be aggravated by the human tendency to use natural sinkholes as dumps in areas of karst topography.

Pollutants and contaminants can be removed from ground water by applying various techniques thereby making it safe for use. Ground water treatment (or remediation) techniques span biological, chemical, and physical treatment technologies. Most ground water treatment techniques utilize a combination of technologies. Some of the biological treatment techniques include bioaugmentation, bioventing, biosparging, bioslurping, and phytoremediation. Some chemical treatment techniques include ozone and oxygen gas injection, chemical precipitation, membrane separation, ion exchange, carbon absorption, aqueous chemical oxidation, and surfactant enhanced recovery. Some chemical techniques may be implemented using nanomaterials. Physical treatment techniques include, but are not limited to, pump and treat, air sparging, and dual phase extraction.

### Abandonment

If treatment or remediation of the polluted groundwater is deemed to be too difficult or expensive then abandoning the use of this aquifer's groundwater and finding an alternative source of water is the only other option

## Light Pollution

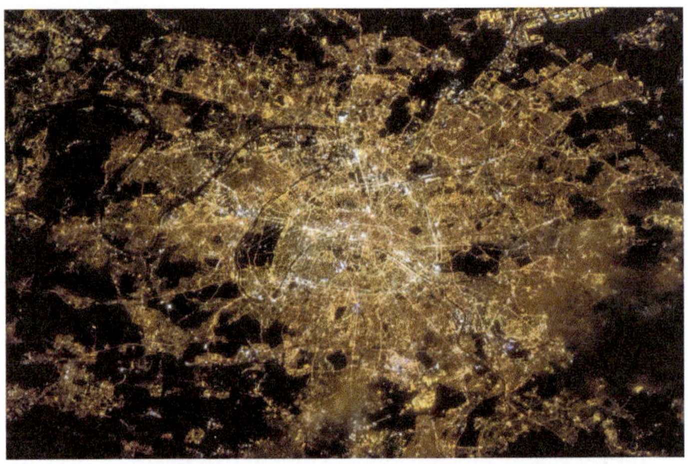

Satellite view of Paris at night (France)

This time exposure photo of New York City at night shows skyglow, one form of light pollution.

A comparison of the view of the night sky from a small rural town (top) and a metropolitan area (bottom). Light pollution dramatically reduces the visibility of stars.

Light pollution, also known as photopollution, is excessive, misdirected or obtrusive artificial light. As a major side-effect of urbanization, it is blamed for compromising health, disrupting ecosystems and spoiling aesthetic environments.

## Definitions

Light pollution is the adding-of/added light itself, in analogy to added sound, carbon dioxide, etc. Adverse consequences are multiple; some of them may not be known yet. Scientific definitions thus include the following:

- The degradation of photic habitat by artificial light.
- The alteration of natural light levels in the outdoor environment owing to artificial light sources.
- The alteration of light levels in the outdoor environment (from those present naturally) due to man-made sources of light. Indoor light pollution is such alteration of light levels in the indoor environment due to sources of light, which compromises human health.
- The introduction by humans, directly or indirectly, of artificial light into the environment.

The first three of the above four scientific definitions describe the state of the environment. The fourth (and newest) one describes the process of polluting by light.

Light pollution competes with starlight in the night sky for urban residents, interferes with astronomical observatories, and, like any other form of pollution, disrupts ecosystems and has adverse health effects.

Light pollution is a side effect of industrial civilization. Its sources include building exterior and interior lighting, advertising, commercial properties, offices, factories, streetlights, and illuminated sporting venues. It is most severe in highly industrialized, densely populated areas of North

America, Europe, and Japan and in major cities in the Middle East and North Africa like Tehran and Cairo, but even relatively small amounts of light can be noticed and create problems. Since the early 1980s, a global dark-sky movement has emerged, with concerned people campaigning to reduce the amount of light pollution. The International Dark-Sky Association (IDA) is one non-profit advocacy group involved in this movement.

## Impact on Energy Usage

Energy conservation advocates contend that light pollution must be addressed by changing the habits of society, so that lighting is used more efficiently, with less waste and less creation of unwanted or unneeded illumination. Several industry groups also recognize light pollution as an important issue. For example, the Institution of Lighting Engineers in the United Kingdom provides its members with information about light pollution, the problems it causes, and how to reduce its impact.

Since not everyone is irritated by the same lighting sources, it is common for one person's light "pollution" to be light that is desirable for another. One example of this is found in advertising, when an advertiser wishes for particular lights to be bright and visible, even though others find them annoying. Other types of light pollution are more certain. For instance, light that *accidentally* crosses a property boundary and annoys a neighbor is generally wasted and pollutive light.

Disputes are still common when deciding appropriate action, and differences in opinion over what light is considered reasonable, and who should be responsible, mean that negotiation must sometimes take place between parties. Where objective measurement is desired, light levels can be quantified by field measurement or mathematical modeling, with results typically displayed as an isophote map or light contour map. Authorities have also taken a variety of measures for dealing with light pollution, depending on the interests, beliefs and understandings of the society involved. Measures range from doing nothing at all, to implementing strict laws and regulations about how lights may be installed and used.

## Types

An example of a light pollution source, using a broad spectrum metal halide lamp, pointing upward at Uniqema factory, Gouda, the Netherlands.

Light pollution is a broad term that refers to multiple problems, all of which are caused by inefficient, unappealing, or (arguably) unnecessary use of artificial light. Specific categories of light pollution include light trespass, over-illumination, glare, light clutter, and skyglow. A single offending light source often falls into more than one of these categories.

## Light Trespass

Light trespass occurs when unwanted light enters one's property, for instance, by shining over a neighbor's fence. A common light trespass problem occurs when a strong light enters the window of one's home from the outside, causing problems such as sleep deprivation. A number of cities in the U.S. have developed standards for outdoor lighting to protect the rights of their citizens against light trespass. To assist them, the International Dark-Sky Association has developed a set of model lighting ordinances.

The Dark-Sky Association was started to reduce the light going up into the sky which reduces visibility of stars. This is any light which is emitted more than 90° above nadir. By limiting light at this 90° mark they have also reduced the light output in the 80–90° range which creates most of the light trespass issues.

U.S. federal agencies may also enforce standards and process complaints within their areas of jurisdiction. For instance, in the case of light trespass by white strobe lighting from communication towers in excess of FAA minimum lighting requirements the Federal Communications Commission maintains an Antenna Structure Registration database information which citizens may use to identify offending structures and provides a mechanism for processing citizen inquiries and complaints. The U.S. Green Building Council (USGBC) has also incorporated a credit for reducing the amount of light trespass and sky glow into their environmentally friendly building standard known as LEED.

Light trespass can be reduced by selecting light fixtures which limit the amount of light emitted more than 80° above the nadir. The IESNA definitions include full cutoff (0%), cutoff (10%), and semi-cutoff (20%). (These definitions also include limits on light emitted above 90° to reduce sky glow.)

## Over-illumination

An office building is illuminated by high pressure sodium (HPS) lamps shining upward, of which much light goes into the sky and neighboring apartment blocks and causes light pollution.

Over-illumination is the excessive use of light. Specifically within the United States, over-illumination is responsible for approximately two million barrels of oil per day in energy wasted. This is based upon U.S. consumption of equivalent of 18.8 million barrels per day (2,990,000 $m^3/d$) of petroleum. It is further noted in the same U.S. Department of Energy source that over 30% of all primary energy is consumed by commercial, industrial and residential sectors. Energy audits of existing buildings demonstrate that the lighting component of residential, commercial and industrial uses consumes about 20–40% of those land uses, variable with region and land use. (Residential use lighting consumes only 10–30% of the energy bill while commercial buildings' major use is lighting.) Thus lighting energy accounts for about four or five million barrels of oil (equivalent) per day. Again energy audit data demonstrates that about 30–60% of energy consumed in lighting is unneeded or gratuitous.

An alternative calculation starts with the fact that commercial building lighting consumes in excess of 81.68 terawatts (1999 data) of electricity, according to the U.S. DOE. Thus commercial lighting alone consumes about four to five million barrels per day (equivalent) of petroleum, in line with the alternate rationale above to estimate U.S. lighting energy consumption. Even among developed countries there are large differences in patterns of light use. American cities emit 3–5 times more light to space per capita compared to German cities.

Over-illumination stems from several factors:

- Not using timers, occupancy sensors or other controls to extinguish lighting when not needed;
- Improper design, especially of workplace spaces, by specifying higher levels of light than needed for a given task;
- Incorrect choice of fixtures or light bulbs, which do not direct light into areas as needed;
- Improper selection of hardware to utilize more energy than needed to accomplish the lighting task;
- Incomplete training of building managers and occupants to use lighting systems efficiently;
- Inadequate lighting maintenance resulting in increased stray light and energy costs;
- "Daylight lighting" demanded by citizens to reduce crime or by shop owners to attract customers;
- Substitution of old mercury lamps with more efficient sodium or metal halide lamps using the same electrical power; and
- Indirect lighting techniques, such as illuminating a vertical wall to bounce light onto the ground.

Most of these issues can be readily corrected with available, inexpensive technology, and with resolution of landlord/tenant practices that create barriers to rapid correction of these matters. Most importantly, public awareness would need to improve for industrialized countries to realize the large payoff in reducing over-illumination.

In certain cases an over-illumination lighting technique may be needed. For example, indirect lighting is often used to obtain a "softer" look, since hard direct lighting is generally found less desirable for certain surfaces, such as skin. The indirect lighting method is perceived as more cozy and suits bars, restaurants and living quarters. It is also possible to block the direct lighting effect by adding softening filters or other solutions, though intensity will be reduced.

## Glare

Glare can be categorized into different types. One such classification is described in a book by Bob Mizon, coordinator for the British Astronomical Association's Campaign for Dark Skies, as follows:

- *Blinding glare* describes effects such as that caused by staring into the Sun. It is completely blinding and leaves temporary or permanent vision deficiencies.
- *Disability glare* describes effects such as being blinded by oncoming car lights, or light scattering in fog or in the eye, reducing contrast, as well as reflections from print and other dark areas that render them bright, with significant reduction in sight capabilities.
- *Discomfort glare* does not typically cause a dangerous situation in itself, though it is annoying and irritating at best. It can potentially cause fatigue if experienced over extended periods.

According to Mario Motta, president of the Massachusetts Medical Society, "... glare from bad lighting is a public-health hazard—especially the older you become. Glare light scattering in the eye causes loss of contrast and leads to unsafe driving conditions, much like the glare on a dirty windshield from low-angle sunlight or the high beams from an oncoming car." In essence bright and/or badly shielded lights around roads can partially blind drivers or pedestrians and contribute to accidents.

The blinding effect is caused in large part by reduced contrast due to light scattering in the eye by excessive brightness, or to reflection of light from dark areas in the field of vision, with luminance similar to the background luminance. This kind of glare is a particular instance of disability glare, called veiling glare. (This is not the same as loss of accommodation of night vision which is caused by the direct effect of the light itself on the eye).

## Light Clutter

The Las Vegas Strip displays excessive groupings of colorful lights. This is a classic example of light clutter.

Light clutter refers to excessive groupings of lights. Groupings of lights may generate confusion, distract from obstacles (including those that they may be intended to illuminate), and potentially cause accidents. Clutter is particularly noticeable on roads where the street lights are badly designed, or where brightly lit advertising surrounds the roadways. Depending on the motives of the person or organization that installed the lights, their placement and design can even be intended to distract drivers, and can contribute to accidents.

Clutter may also present a hazard in the aviation environment if aviation safety lighting must compete for pilot attention with non-relevant lighting. For instance, runway lighting may be confused with an array of suburban commercial lighting and aircraft collision avoidance lights may be confused with ground lights.

## Skyglow

Mexico City at night, with a brightly illuminated sky

In pristine areas, clouds appear black and blot out the stars. In urban areas, clouds strongly enhance skyglow.

Skyglow refers to the diffuse glow that can be seen over populated areas. It arises from light reflected from illuminated surfaces and from light escaping directly upward from incompletely shielded or upward-directed light fixtures, which then is scattered (redirected) by the atmosphere back toward the ground. The brightness of skyglow is affected strongly by the amount of light used, the shielding characteristics of the light fixtures, and by the color or spectral content of the light sources. Though it is commonly thought that Rayleigh scattering strongly increases the brightness of skyglow arising from white sources with strong blue and green emission, the dominant spectral effect has been shown to arise from the Purkinje effect instead. Because of the eye's increased sensitivity to blue and green light when adapted to very low luminance levels (such as those in even moderately polluted clear night skies), white light sources such as white LEDs contribute significantly more to skyglow than an equal amount of yellow light from sources such as high-pressure sodium, low-pressure sodium, "phosphor-converted" (PC) amber LED, or "narrow spectrum" amber AlGaInP LEDs. Sky glow is of particular irritation to astronomers, as it interferes with astronomical observations, but multitudes suffer from the decreased visibility of star-filled night skies.

Skyglow brightness can be measured with instruments such as the Sky Quality Meters (SQM), or visually using the Bortle Dark-Sky Scale. The nine-class Bortle Scale rates the darkness of the night sky and the visibility of its phenomena, such as the Milky Way, gegenschein and the zodiacal light (easily masked by skyglow), providing a detailed description of each level on the scale (with Class 1 being the best).

Light is particularly problematic for amateur astronomers, whose ability to observe the night sky from their property is inhibited by any stray light from nearby. Most major optical astronomical observatories are surrounded by zones of strictly enforced restrictions on light emissions to limit skyglow.

Direct skyglow is reduced by selecting lighting fixtures which limit the amount of light emitted above the horizontal (90° above the nadir). Indirect skyglow produced by reflections from illuminated surfaces is harder to manage; effective methods for limiting it include the use of low-impact lamp types (e.g. amber sources), minimizing over-illumination and inefficient designs or fixtures that spread light where it is not needed. But it has to be taken into account that, according to late 2010 publications, Italian regions using full cut off lighting only does not increase skyglow. Anyway, light reflected upwards by dark surfaces such as roads or buildings can be considered as minor, so debate about the contribution of indirect skyglow will continue.

Skyglow is made considerably worse when clouds are present. While this has no effect on astronomical observations (which are not possible at visible wavelengths under cloud cover), it is very important in the context of ecological light pollution. Since cloudy nights in artificially lit areas can be up to ten thousand times brighter than in natural areas, any organisms that are affected by sky glow (e.g. zooplankton and fish that visually prey on them) are much more likely to have their ordinary behavior disturbed on cloudy nights.

Impacts of skyglow are explored in the photography and timelapse project SKYGLOW, created in collaboration with International Dark-Sky Association.

## Measurement and Global Effects

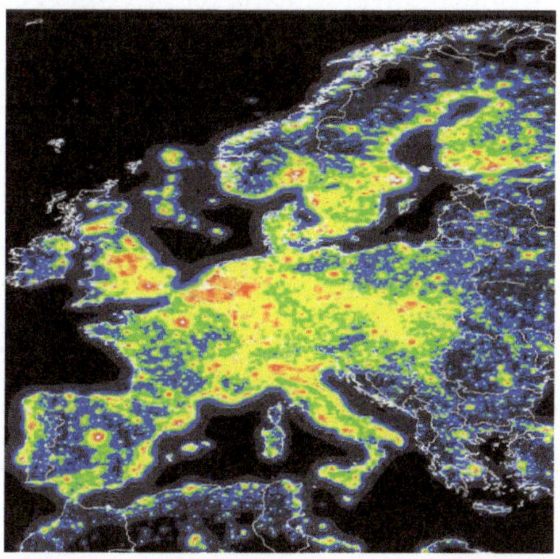

False colors show intensities of skyglow from artificial light sources.

Measuring the effect of sky glow on a global scale is a complex procedure. The natural atmosphere is not completely dark, even in the absence of terrestrial sources of light and illumination from the Moon. This is caused by two main sources: *airglow* and *scattered light*.

At high altitudes, primarily above the mesosphere, there is enough UV radiation from the sun of very short wavelength to cause ionization. When the ions collide with electrically neutral particles they recombine and emit photons in the process, causing airglow. The degree of ionization is sufficiently large to allow a constant emission of radiation even during the night when the upper atmosphere is in the Earth's shadow. Lower in the atmosphere all of the solar photons with energies above the ionization potential of $N_2$ and $O_2$ have already been absorbed by the higher layers and thus no appreciable ionization occurs.

Apart from emitting light, the sky also scatters incoming light, primarily from distant stars and the Milky Way, but also the zodiacal light, sunlight that is reflected and backscattered from interplanetary dust particles.

The amount of airglow and zodiacal light is quite variable (depending, amongst other things on sunspot activity and the Solar cycle) but given optimal conditions the darkest possible sky has a brightness of about 22 magnitude/square arcsecond. If a full moon is present, the sky brightness increases to about 18 magnitude/sq. arcsecond depending on local atmospheric transparency, 40 times brighter than the darkest sky. In densely populated areas a sky brightness of 17 magnitude/sq. arcsecond is not uncommon, or as much as 100 times brighter than is natural.

To precisely measure how bright the sky gets, night time satellite imagery of the earth is used as raw input for the number and intensity of light sources. These are put into a physical model of scattering due to air molecules and aerosoles to calculate cumulative sky brightness. Maps that show the enhanced sky brightness have been prepared for the entire world.

Inspection of the area surrounding Madrid reveals that the effects of light pollution caused by a single large conglomeration can be felt up to 100 km (62 mi) away from the center. Global effects of light pollution are also made obvious. The entire area consisting of southern England, Netherlands, Belgium, west Germany, and northern France have a sky brightness of at least 2 to 4 times above normal. The only places in continental Europe where the sky can attain its natural darkness are in northern Scandinavia and in islands far from the continent.

In North America the situation is comparable. There is a significant problem with light pollution ranging from the Canadian Maritime Provinces to the American Southwest. The International Dark-Sky Association works to designate areas that have high quality night skies. These areas are supported by communities and organizations that are dedicated to reducing light pollution (e.g.Dark-sky preserve). The National Park Service Natural Sounds and Night Skies Division has measured night sky quality in national park units across the U.S. Sky quality in the U.S. ranges from pristine (Capitol Reef National Park and Big Bend National Park) to severely degraded (Santa Monica Mountains National Recreation Area and Biscayne National Park). The National Park Service Night Sky Program monitoring database is available online (2015).

Light pollution in Hong Kong was declared the 'worst on the planet' in March 2013.

In June 2016, it was estimated that one third of the world's population could no longer see the Milky Way, including 80% of Americans and 60% of Europeans. Singapore was found to be the most light-polluted country in the world.

A worldwide cloud-free mosaic from the Suomi NPP satellite, showing the extent of visible lights in 2016. The effects of light pollution – especially skyglow – spread far beyond the light sources visible here.

## Consequences

### Energy Waste

Christmas lights in California.

Lighting is responsible for one-fourth of all electricity consumption worldwide, and case studies have shown that several forms of over-illumination constitute energy wastage, including non-beneficial upward direction of night-time lighting. In 2007, Terna, the company responsible for managing electricity flow in Italy, reported a saving of 645.2 million kWh in electricity consumption during the daylight saving period from April to October. It attributes this saving to the delayed need for artificial lighting during the evenings.

In Australia,

... public lighting is the single largest source of local government's greenhouse gas emissions, typically accounting for 30 to 50% of their emissions. There are 1.94 million public lights — one for every 10 Australians — that annually cost A$210 million, use 1,035 GWh of electricity and are responsible for 1.15 million tonnes of CO2 emissions.

Current public lighting in Australia, particularly for minor roads and streets, uses large amounts of energy and financial resources, while often failing to provide high quality lighting. There are many ways to improve lighting quality while reducing energy use and greenhouse gas emissions as well as lowering costs.

## Effects on Animal and Human Health and Psychology

Streetlights in the skiing resort Kastelruth in South Tyrol.

Medical research on the effects of excessive light on the human body suggests that a variety of adverse health effects may be caused by light pollution or excessive light exposure, and some lighting design textbooks use human health as an explicit criterion for proper interior lighting. Health effects of over-illumination or improper spectral composition of light may include: increased headache incidence, worker fatigue, medically defined stress, decrease in sexual function and increase in anxiety. Likewise, animal models have been studied demonstrating unavoidable light to produce adverse effect on mood and anxiety. For those who need to be awake at night, light at night also has an acute effect on alertness and mood.

In 2007, "shift work that involves circadian disruption" was listed as a probable carcinogen by the World Health Organization's International Agency for Research on Cancer. (IARC Press release No. 180). Multiple studies have documented a correlation between night shift work and the increased incidence of breast and prostate cancer.

A more recent discussion (2009), written by Professor Steven Lockley, Harvard Medical School, can be found in the CfDS handbook "Blinded by the Light?". "Human health implications of light pollution" states that "... light intrusion, even if dim, is likely to have measurable effects on sleep disruption and melatonin suppression. Even if these effects are relatively small from night to night, continuous chronic circadian, sleep and hormonal disruption may have longer-term health risks". The New York Academy of Sciences hosted a meeting in 2009 on Circadian Disruption and Cancer. Red light suppresses melatonin the least.

In June 2009, the American Medical Association developed a policy in support of control of light pollution. News about the decision emphasized glare as a public health hazard leading to unsafe driving conditions. Especially in the elderly, glare produces loss of contrast, obscuring night vision.

## Disruption of Ecosystems

When artificial light affects organisms and ecosystems it is called ecological light pollution. While light at night can be beneficial, neutral, or damaging for individual species, its presence invariably disturbs ecosystems. For example, some species of spiders avoid lit areas, while other species are happy to build their spider web directly on a lamp post. Since lamp posts attract many flying insects, the spiders that don't mind light gain an advantage over the spiders that avoid it. This is a simple example of the way in which species frequencies and food webs can be disturbed by the introduction of light at night.

Light pollution poses a serious threat in particular to nocturnal wildlife, having negative impacts on plant and animal physiology. It can confuse animal navigation, alter competitive interactions, change predator-prey relations, and cause physiological harm. The rhythm of life is orchestrated by the natural diurnal patterns of light and dark, so disruption to these patterns impacts the ecological dynamics.

Studies suggest that light pollution around lakes prevents zooplankton, such as Daphnia, from eating surface algae, causing algal blooms that can kill off the lakes' plants and lower water quality. Light pollution may also affect ecosystems in other ways. For example, lepidopterists and entomologists have documented that nighttime light may interfere with the ability of moths and other nocturnal insects to navigate. Night-blooming flowers that depend on moths for pollination may be affected by night lighting, as there is no replacement pollinator that would not be affected by the artificial light. This can lead to species decline of plants that are unable to reproduce, and change an area's longterm ecology.

A 2009 study also suggests deleterious impacts on animals and ecosystems because of perturbation of polarized light or artificial polarization of light (even during the day, because direction of natural polarization of sun light and its reflexion is a source of information for a lot of animals). This form of pollution is named polarized light pollution (PLP). Unnatural polarized light sources can trigger maladaptive behaviors in polarization-sensitive taxa and alter ecological interactions.

Lights on tall structures can disorient migrating birds. Estimates by the U.S. Fish and Wildlife Service of the number of birds killed after being attracted to tall towers range from 4 to 5 million per year to an order of magnitude higher. The Fatal Light Awareness Program (FLAP) works with building owners in Toronto, Canada and other cities to reduce mortality of birds by turning out lights during migration periods.

Similar disorientation has also been noted for bird species migrating close to offshore production and drilling facilities. Studies carried out by Nederlandse Aardolie Maatschappij b.v. (NAM) and Shell have led to development and trial of new lighting technologies in the North Sea. In early 2007, the lights were installed on the Shell production platform L15. The experiment proved a great success since the number of birds circling the platform declined by 50 to 90%.

Sea turtle hatchlings emerging from nests on beaches are another casualty of light pollution. It is a common misconception that hatchling sea turtles are attracted to the moon. Rather, they find the ocean by moving away from the dark silhouette of dunes and their vegetation, a behavior with which artificial lights interfere. The breeding activity and reproductive phenology of toads, however, are cued by moonlight. Juvenile seabirds may also be disoriented by lights as they leave their nests and fly out to sea. Amphibians and reptiles are also affected by light pollution. Introduced light sources during normally dark periods can disrupt levels of melatonin production. Melatonin is a hormone that regulates photoperiodic physiology and behaviour. Some species of frogs and salamanders utilize a light-dependent "compass" to orient their migratory behaviour to breeding sites. Introduced light can also cause developmental irregularities, such as retinal damage, reduced sperm production, and genetic mutation.

In September 2009, the 9th European Dark-Sky Symposium in Armagh, Northern Ireland had a session on the environmental effects of light at night (LAN). It dealt with bats, turtles, the "hidden" harms of LAN, and many other topics. The environmental effects of LAN were mentioned as early as 1897, in a *Los Angeles Times* article. The following is an excerpt from that article, called "Electricity and English songbirds".

An English journal has become alarmed at the relation of electricity to songbirds, which it maintains is closer than that of cats and fodder crops. How many of us, it asks, foresee that electricity may extirpate the songbird?...With the exception of the finches, all the English songbirds may be said to be insectivorous, and their diet consists chiefly of vast numbers of very small insects which they collect from the grass and herbs before the dew is dry. As the electric light is finding its way for street illumination into the country parts of England, these poor winged atoms are slain by thousands at each light every warm summer evening... The fear is expressed, that when England is lighted from one end to the other with electricity the song birds will die out from the failure of their food supply.

## Effect on Astronomy

The constellation Orion, imaged at left from dark skies, and at right from within the Provo/Orem, Utah metropolitan area.

Astronomy is very sensitive to light pollution. The night sky viewed from a city bears no resemblance to what can be seen from dark skies. Skyglow (the scattering of light in the atmosphere) reduces the contrast between stars and galaxies and the sky itself, making it much harder to see fainter objects. This is one factor that has caused newer telescopes to be built in increasingly remote areas. Some astronomers use narrow-band "nebula filters" which only allow specific wavelengths of light commonly seen in nebulae, or broad-band "light pollution filters" which are designed to reduce (but not eliminate) the effects of light pollution by filtering out spectral lines commonly emitted by sodium- and mercury-vapor lamps, thus enhancing contrast and improving the view of dim objects such as galaxies and nebulae. Unfortunately these light pollution reduction (LPR) filters are not a cure for light pollution. LPR filters reduce the brightness of the object under study and this limits the use of higher magnifications. LPR filters work by blocking light of certain wavelengths, which alters the color of the object, often creating a pronounced green cast. Furthermore, LPR filters only work on certain object types (mainly emission nebulae) and are of little use on galaxies and stars. No filter can match the effectiveness of a dark sky for visual or photographic purposes. Due to their low surface brightness, the visibility of diffuse sky objects such as nebulae and galaxies is affected by light pollution more than are stars. Most such objects are rendered invisible in heavily light polluted skies around major cities. A simple method for estimating the darkness of a location is to look for the Milky Way, which from truly dark skies appears bright enough to cast a shadow.

Outskirts of the Atacama Desert, far from the light-polluted cities of northern Chile, the skies are pitch-black after sunset.

In addition to skyglow, light trespass can impact observations when artificial light directly enters the tube of the telescope and is reflected from non-optical surfaces until it eventually reaches the eyepiece. This direct form of light pollution causes a glow across the field of view which reduces contrast. Light trespass also makes it hard for a visual observer to become sufficiently dark adapted. The usual measures to reduce this glare, if reducing the light directly is not an option, include flocking the telescope tube and accessories to reduce reflection, and putting a light shield (also usable as a dew shield) on the telescope to reduce light entering from angles other than those near the target. Under these conditions, some astronomers prefer to observe under a black cloth to ensure maximum dark adaptation. In one Italian regional lighting code this effect of stray light is defined as "optical pollution", duc to the fact that there is a direct path from the light source to the "optic" – the observer's eye or telescope.

## Increase in Atmospheric Pollution

A study presented at the American Geophysical Union meeting in San Francisco found that light pollution destroys nitrate radicals thus preventing the normal night time reduction of atmospheric smog produced by fumes emitted from cars and factories. The study was presented by Harald Stark from the National Oceanic and Atmospheric Administration.

## Reduction of Natural Sky Polarization

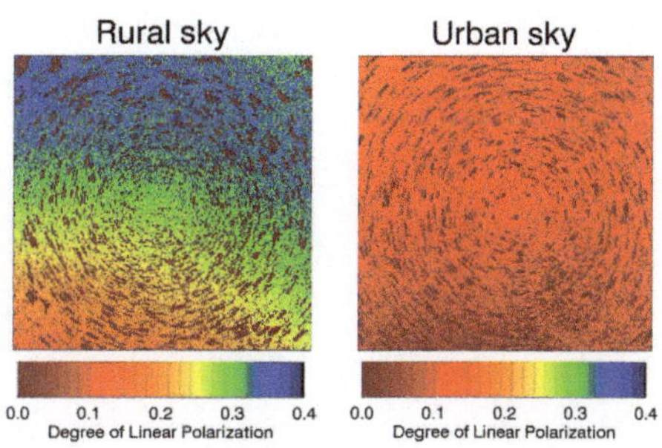


Light pollution is mostly unpolarized, and its addition to moonlight results in a decreased polarization signal.

In the night, the polarization of the moonlit sky is very strongly reduced in the presence of urban light pollution, because scattered urban light is not strongly polarized. Polarized moonlight can't be seen by humans, but is believed to be used by many animals for navigation.

## Reduction

This kind of LED droplight could reduce unnecessary light pollution in building interiors

Reducing light pollution implies many things, such as reducing sky glow, reducing glare, reducing light trespass, and reducing clutter. The method for best reducing light pollution, therefore, depends on exactly what the problem is in any given instance. Possible solutions include:

- Utilizing light sources of minimum intensity necessary to accomplish the light's purpose.
- Turning lights off using a timer or occupancy sensor or manually when not needed.
- Improving lighting fixtures, so that they direct their light more accurately towards where it is needed, and with fewer side effects.
- Adjusting the *type* of lights used, so that the light waves emitted are those that are less likely to cause severe light pollution problems. Mercury, metal halide and above all first generation of blue-light LED road luminaires are much more pollutant than sodium lamps: Earth atmosphere scatters and transmits blue light better than yellow or red light. It is a common experience observing "glare" and "fog" around and below LED road luminaires as soon as air humidity increases, while orange sodium lamp luminaires are less prone to show this phenomenon.
- Evaluating existing lighting plans, and re-designing some or all of the plans depending on whether existing light is actually needed.

## Improving Lighting Fixtures

The use of *full cutoff* lighting fixtures, as much as possible, is advocated by most campaigners for the reduction of light pollution. It is also commonly recommended that lights be spaced appropriately for maximum efficiency, and that number of luminaires being used as well as the wattage of each luminaire match the needs of the particular application (based on local lighting design standards).

A *flat-lens cobra luminaire*, which is a full-cutoff fixture, is very effective in reducing light pollution. It ensures that light is only directed below the horizontal, which means less light is wasted through directing it outwards and upwards.

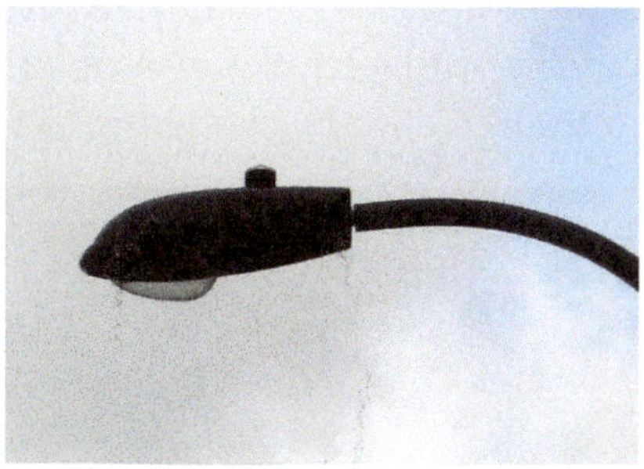

This *drop-lens cobra luminaire* allows light to escape sideways and upwards, where it may cause problems.

Full cutoff fixtures first became available in 1959 with the introduction of General Electric's M100 fixture.

A full cutoff fixture, when correctly installed, reduces the chance for light to escape above the plane of the horizontal. Light released above the horizontal may sometimes be lighting an intended target, but often serves no purpose. When it enters into the atmosphere, light contributes to sky glow. Some governments and organizations are now considering, or have already implemented, full cutoff fixtures in street lamps and stadium lighting.

The use of full cutoff fixtures help to reduce sky glow by preventing light from escaping above the horizontal. Full cutoff typically reduces the visibility of the lamp and reflector within a luminaire, so the effects of glare are also reduced. Campaigners also commonly argue that full cutoff fixtures are more efficient than other fixtures, since light that would otherwise have escaped into the atmosphere may instead be directed towards the ground. However, full cutoff fixtures may also trap more light in the fixture than other types of luminaires, corresponding to lower luminaire efficiency, suggesting a re-design of some luminaires may be necessary.

The use of full cutoff fixtures can allow for lower wattage lamps to be used in the fixtures, producing the same or sometimes a better effect, due to being more carefully controlled. In every lighting system, some sky glow also results from light reflected from the ground. This reflection can be reduced, however, by being careful to use only the lowest wattage necessary for the lamp, and setting spacing between lights appropriately. Assuring luminaire setback is greater than 90° from highly reflective surfaces also diminishes reflectance.

A common criticism of full cutoff lighting fixtures is that they are sometimes not as aesthetically pleasing to look at. This is most likely because historically there has not been a large market specifically for full cutoff fixtures, and because people typically like to see the source of illumination. Due to the specificity with their direction of light, full cutoff fixtures sometimes also require expertise to install for maximum effect.

The effectiveness of using full cutoff roadway lights to combat light pollution has also been called into question. According to design investigations, luminaires with full cutoff distributions (as opposed to *cutoff* or *semi cutoff*, compared here) have to be closer together to meet the same light level, uniformity and glare requirements specified by the IESNA. These simulations optimized the height and spacing of the lights while constraining the overall design to meet the IESNA requirements, and then compared total uplight and energy consumption of different luminaire designs and powers. Cutoff designs performed better than full cutoff designs, and semi-cutoff performed better than either cutoff or full cutoff. This indicates that, in roadway installations, over-illumination or poor uniformity produced by full cutoff fixtures may be more detrimental than direct uplight created by fewer cutoff or semi-cutoff fixtures. Therefore, the overall performance of existing systems could be improved more by reducing the number of luminaires than by switching to full cutoff designs.

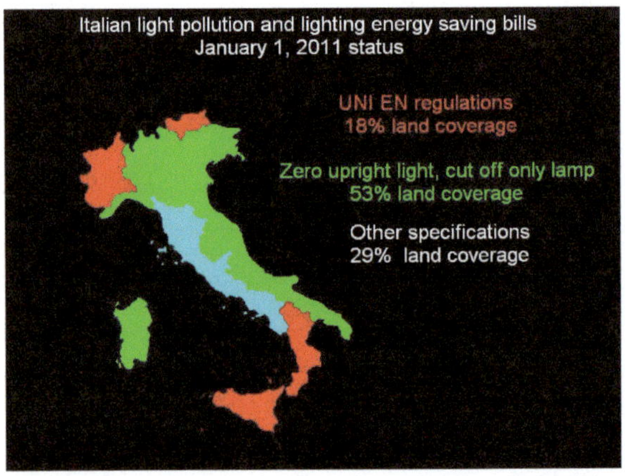


The majority of Italian regions require "zero upward light", which usually implies use of overall full cut-off lamps for new luminaires, but violations are common.

However, using the definition of "light pollution" from some Italian regional bills (i.e., "every irradiance of artificial light outside competence areas and particularly upward the sky") only full cutoff design prevents light pollution. The Italian Lombardy region, where only full cutoff design is allowed (Lombardy act no. 17/2000, promoted by Cielobuio-coordination for the protection of the night sky), in 2007 had the lowest per capita energy consumption for public lighting in Italy. The same legislation also imposes a minimum distance between street lamps of about four times their height, so full cut off street lamps are the best solution to reduce both light pollution and electrical power usage.

## Adjusting Types of Light Sources

Several different types of light sources exist, each having different properties that affect their appropriateness for certain tasks, particularly efficiency and spectral power distribution. It is often the case that inappropriate light sources have been selected for a task, either due to ignorance or because more sophisticated light sources were unavailable at the time of installation. Therefore, badly chosen light sources often contribute unnecessarily to light pollution and energy waste. By re-assessing and changing the light sources used, it is often possible to reduce energy use and pollutive effects while simultaneously greatly improving efficiency and visibility.

Some types of light sources are listed in order of energy efficiency in the table below (figures are approximate maintained values), and include relative visual skyglow impacts.

| Type of light source | Color | Luminous efficiency (in lumens per watt) | Sky glow impact (relative to LPS) |
|---|---|---|---|
| LED street light (white) | warm-white to cool-white | 120 | 4–8 |
| Low Pressure Sodium (LPS/SOX) | yellow/amber | 110 | 1.0 |
| High Pressure Sodium (HPS/SON) | pink/amber-white | 90 | 2.4 |
| Metal Halide | warm-white to cool-white | 70 | 4–8 |
| Incandescent | yellow/white | 8–25 | 1.1 |

Many astronomers request that nearby communities use low pressure sodium lights or amber Aluminium gallium indium phosphide LED as much as possible, because the principal wavelength emitted is comparably easy to work around or in rare cases filter out. The low cost of operating sodium lights is another feature. In 1980, for example, San Jose, California, replaced all street lamps with low pressure sodium lamps, whose light is easier for nearby Lick Observatory to filter out. Similar programs are now in place in Arizona and Hawaii. Such yellow light sources also have significantly less visual skyglow impact, so reduce visual sky brightness and improve star visibility for everyone.

Disadvantages of low pressure sodium lighting are that fixtures must usually be larger than competing fixtures, and that color cannot be distinguished, due to its emitting principally a single wavelength of light. Due to the substantial size of the lamp, particularly in higher wattages such as 135 W and 180 W, control of light emissions from low pressure sodium luminaires is more difficult. For applications requiring more precise direction of light (such as narrow roadways) the native lamp efficacy advantage of this lamp type is decreased and may be entirely lost compared to high pressure sodium lamps. Allegations that this also leads to higher amounts of light pollution from luminaires running these lamps arise principally because of older luminaires with poor shielding, still widely in use in the UK and in some other locations. Modern low-pressure sodium fixtures with better optics and full shielding, and the decreased skyglow impacts of yellow light preserve the luminous efficacy advantage of low-pressure sodium and result in most cases is less energy consumption and less visible light pollution. Unfortunately, due to continued lack of accurate information, many lighting professionals continue to disparage low-pressure sodium, contributing to its decreased acceptance and specification in lighting standards and therefore its use. Another disadvantage of low-pressure sodium lamps is that some people find the characteristic yellow light very displeasing aesthetically.

Because of the increased sensitivity of the human eye to blue and green wavelengths when viewing low-luminances (the Purkinje effect) in the night sky, different sources produce dramatically different amounts of visible skyglow from the same amount of light sent into the atmosphere.

## Re-designing Lighting Plans

In some cases, evaluation of existing plans has determined that more efficient lighting plans are possible. For instance, light pollution can be reduced by turning off unneeded outdoor lights, and only lighting stadiums when there are people inside. Timers are especially valuable for this purpose. One of the world's first coordinated *legislative* efforts to reduce the adverse effect of this pollution on the environment began in Flagstaff, Arizona, in the U.S. There, over three decades of ordinance development has taken place, with the full support of the population, often with government support, with community advocates, and with the help of major local observatories, including the United States Naval Observatory Flagstaff Station. Each component helps to educate, protect and enforce the imperatives to intelligently reduce detrimental light pollution.

One example of a lighting plan assessment can be seen in a report originally commissioned by the Office of the Deputy Prime Minister in the United Kingdom, and now available through the Department for Communities and Local Government. The report details a plan to be implemented throughout the UK, for designing lighting schemes in the countryside, with a particular focus on preserving the environment.

In another example, the city of Calgary has recently replaced most residential street lights with models that are comparably energy efficient. The motivation is primarily operation cost and environmental conservation. The costs of installation are expected to be regained through energy savings within six to seven years.

Crossroad in Alessandria, Italy: luminaires with mercury lamps are in the background, LED street lights in the middle, luminaires with high pressure sodium lamps are in the foreground.

The Swiss Agency for Energy Efficiency (SAFE) uses a concept that promises to be of great use in the diagnosis and design of road lighting, "*consommation électrique spécifique* (*CES*)", which can be translated into English as "specific electric power consumption (SEC)". Thus, based on observed lighting levels in a wide range of Swiss towns, SAFE has defined target values for electric power consumption per metre for roads of various categories. Thus, SAFE currently recommends an SEC of 2 to 3 watts per meter for roads of less than 10 metre width (4 to 6 watts per metre for wider roads). Such a measure provides an easily applicable environmental protection constraint on conventional "norms", which usually are based on the recommendations of lighting manufacturing interests, who may not take into account environmental criteria. In view of ongoing progress in lighting technology, target SEC values will need to be periodically revised downwards.

A newer method for predicting and measuring various aspects of light pollution was described in the journal Lighting Research Technology (September 2008). Scientists at Rensselaer Polytechnic Institute's Lighting Research Center have developed a comprehensive method called Outdoor Site-Lighting Performance (OSP), which allows users to quantify, and thus optimize, the performance of existing and planned lighting designs and applications to minimize excessive or obtrusive light leaving the boundaries of a property. OSP can be used by lighting engineers immediately, particularly for the investigation of glow and trespass (glare analyses are more complex to perform and current commercial software does not readily allow them), and can help users compare several lighting design alternatives for the same site.

In the effort to reduce light pollution, researchers have developed a "Unified System of Photometry," which is a way to measure how much or what kind of street lighting is needed. The Unified System of Photometry allows light fixtures to be designed to reduce energy use while maintaining or improving perceptions of visibility, safety, and security. There was a need to create a new system of light measurement at night because the biological way in which the eye's rods and cones process light is different in nighttime conditions versus daytime conditions. Using this new system of photometry, results from recent studies have indicated that replacing traditional, yellowish, high-pressure sodium (HPS) lights with "cool" white light sources, such as induction, fluorescent, ceramic metal halide, or LEDs can actually reduce the amount of electric power used for lighting while maintaining or improving visibility in nighttime conditions.

The International Commission on Illumination, also known as the CIE from its French title, la Commission Internationale de l'Eclairage, will soon be releasing its own form of unified photometry for outdoor lighting.

## Neuroplastic Effects of Pollution

Research indicates that living in areas of high pollution has serious long term health effects. Living in these areas during childhood and adolescence can lead to diminished mental capacity and an increased risk of brain damage. People of all ages who live in high pollution areas for extended periods place themselves at increased risk of various neurological disorders. Both air pollution and heavy metal pollution have been implicated as having negative effects on central nervous system (CNS) functionality. The ability of pollutants to affect the neurophysiology of individuals after the structure of the CNS has become mostly stabilized is an example of negative neuroplasticity.

### Air Pollution

Air pollution is known to affect small and large blood vessels throughout the body. High levels of air pollution are associated with increased risk of strokes and heart attacks. By permanently affecting vascular structures in the brain, air pollution can have serious effects on neural functioning and neural matter. In dogs air pollution shows to cause damage to the CNS by altering the blood–brain barrier, causing neurons in the cerebral cortex to degenerate, destroying glial cells found in white matter, and by causing neurofibrillary tangles. These changes can permanently alter brain structure and chemistry, resulting in various impairments and disorders. Sometimes, the effects of neural remodeling do not manifest themselves for a prolonged period of time.

## Effects in Adolescents and Canines

A study from 2008 compared children and dogs raised in Mexico City (a location known for high pollution levels) with children and dogs raised in Polotitlán, Mexico (a city whose pollution levels meet the current US National Ambient Air Quality Standards). According to this study, children raised in areas of higher pollution scored lower in intelligence (i.e. on IQ tests), and showed signs of lesions in MRI scanning of the brain. In contrast, children from the low pollution area scored as expected on IQ tests, and did not show any significant sign of the risk of brain lesions. This correlation was found to be statistically significant, and shows that pollution levels may be related to, and contribute to, brain lesion formation and IQ scores, which, in turn, manifests as impaired intellectual capacity and/or performance. Living in high pollution areas thus places adolescents at risk of premature brain degeneration and improper neural development—these findings could have significant implications for future generations.

## Effects in Adults

There are indications that the effects of physical activity and air pollution on neuroplasticity counteract. Physical activity is known for its health-enhancing benefits, particularly on the cardiovascular system, and has also demonstrated benefits for brain plasticity processes, cognition and mental health. The neurotrophine, brain-derived neurotrophic factor (BDNF) is thought to play a key role in exercise-induced cognitive improvements. Brief bouts of physical activity have been shown to increase serum levels of BDNF, but this increase may be offset by increased exposure to traffic-related air pollution. Over longer periods of physical exercise, cognitive improvements that were demonstrated in rural joggers were found to be absent in urban joggers taking the same 12-week start-2-run training programme.

## Epilepsy

Researchers in Chile found statistically-significant correlations between multiple air pollutants and the risk of epilepsy using a 95% confidence interval. The air pollutants that the researchers attempted to correlate with increased incidence of epilepsy included carbon monoxide, ozone, sulfur dioxide, nitrogen dioxide, large particulate matter, and fine particulate matter. The researchers tested these pollutants across seven cities and, in all but one case, a correlation was found between pollutant levels and the occurrence of epilepsy. Interestingly, all of the correlations found were shown to be statistically significant. The researchers hypothesized that air pollutants increase epilepsy risk by increasing inflammatory mediators, and by providing a source of oxidative stress. They believe that these changes eventually alter the functioning of the blood–brain barrier, causing brain inflammation. Brain inflammation is known to be a risk factor for epilepsy; thus, the sequence of events provides a plausible mechanism by which pollution may increase epilepsy risk in individuals who are genetically vulnerable to the disease.

## Dioxin Poisoning

Organohalogen compounds, such as dioxins, are commonly found in pesticides or created as by-products of pesticide manufacture or degradation. These compounds can have a significant impact on the neurobiology of exposed organisms. Some observed effects of exposure to dioxins are altered astroglial intracellular calcium ion ($Ca^{2+}$), decreased glutathione levels, modified neurotransmitter function in the CNS, and loss of pH maintenance. A study of 350 chemical plant employees exposed to a dioxin precursor for herbicide synthesis between 1965 and 1968 showed that 80 of the employees displayed signs of dioxin poisoning. Of these 350 employees, 15 were contacted again in 2004 to submit to neurological tests to assess whether the dioxin poisoning had any long-term effects on neurological capabilities. The amount of time that had passed made it difficult to assemble a larger cohort, but the results of the tests indicated that eight of the 15 subjects exhibited some central nervous system impairment, nine showed signs of polyneuropathy, and electroencephalography (EEG) showed various degrees of structural abnormalities. This study suggested that the effects of dioxins were not limited to initial toxicity. Dioxins, through neuroplastic effects, can cause long-term damage that may not manifest itself for years or even decades.

## Metal Exposure

Heavy metal exposure can result in an increased risk of various neurological diseases. Research indicates that the two most neurotoxic heavy metals are mercury and lead. The impact that these two metals will have is highly dependent upon the individual due to genetic variations. Mercury and lead are particularly neurotoxic for many reasons: they easily cross cell membranes, have oxidative effects on cells, react with sulfur in the body (leading to disturbances in the many functions that rely upon sulfhydryl groups), and reduce glutathione levels inside cells. Methylmercury, in particular, has an extremely high affinity for sulfhydryl groups. Organomercury is a particularly damaging form of mercury because of its high absorbability Lead also mimics calcium, a very important mineral in the CNS, and this mimicry leads to many adverse effects. Mercury's neuroplastic mechanisms work by affecting protein production. Elevated mercury levels increase glutathione levels by affecting gene expression, and this in turn affects two proteins (MT1 and MT2) that are contained in astrocytes and neurons. Lead's ability to imitate calcium allows it to cross the blood–brain barrier. Lead also upregulates glutathione.

## Autism

Heavy metal exposure, when combined with certain genetic predispositions, can place individuals at increased risk for developing autism. Many examples of CNS pathophysiology, such as oxidative stress, neuroinflammation, and mitochondrial dysfunction, could be by-products of environmental stressors such as pollution. There have been reports of autism outbreaks occurring in specific locations. Since these cases of autism are related to geographic location, the implication is that something in the environment is complementing an at-risk genotype to cause autism in these vulnerable individuals. Mercury and lead both contribute to inflammation, leading scientists to speculate that these heavy metals could play a role in autism. These findings are controversial, however, with many researchers believing that increasing rates of autism are a consequence of more accurate screening and diagnostic methods, and are not due to any sort of environmental factor.

## Accelerated Neural Aging

Neuroinflammation is associated with increased rates of neurodegeneration. Inflammation tends to increase naturally with age. By facilitating inflammation, pollutants such as air particulates and heavy metals cause the CNS to age more quickly. Many late-onset diseases are caused by neurodegeneration. Multiple sclerosis, Parkinson's disease, amyotrophic lateral sclerosis (ALS), and Alzheimer's disease are all believed to be exacerbated by inflammatory processes, resulting in individuals displaying signs of these diseases at an earlier age than is typically expected.

Multiple sclerosis occurs when chronic inflammation leads to the compromise of oligodendrocytes, which in turn leads to the destruction of the myelin sheath. Then axons begin exhibiting signs of damage, which in turn leads to neuron death. Multiple sclerosis has been correlated to living in areas with high particulate matter levels in the air.

In Parkinson's disease, inflammation leading to depletion of antioxidant stores will ultimately lead to dopaminergic neuron degeneration, causing a shortage of dopamine and contributing to the formation of Parkinson's disease. Chronic glial activation as a result of inflammation causes motor neuron death and compromises astrocytes, these factors leading to the symptoms of amyotrophic lateral sclerosis (ALS, aka Lou Gehrig's disease).

In the case of Alzheimer's disease, inflammatory processes lead to neuron death by inhibiting growth at axons and activating astrocytes that produce proteoglycans. This product can only be deposited in the hippocampus and cortex, indicating that this may be the reason these two areas show the highest levels of degeneration in Alzheimer's disease. Airborne metal particulates have been shown to directly access and affect the brain through olfactory pathways, which allows a large amount of particulate matter to reach the blood–brain barrier.

These facts, coupled with air pollution's link to neurofibrillary tangles and the observed subcortical vascular changes observed in dogs, imply that the negative neuroplastic effects of pollution could result in increased risk for Alzheimer's disease, and could also implicate pollution as a cause of early-onset Alzheimer's disease through multiple mechanisms. The general effect of pollution is increased levels of inflammation. As a result, pollution can significantly contribute to various neurological disorders that are caused by inflammatory processes.

## Toxic Hotspot

Toxic hotspots are locations where emissions from specific sources such as water or air pollution may expose local populations to elevated health risks, such as cancer. These emissions contribute to cumulative health risks of emissions from other sources nearby. Urban, highly populated areas around pollutant emitters such as old factories and waste storage sites are often toxic hotspots.

### Soil Contamination Hotspots

The 1984 Bhopal disaster in India, the world's worst chemical disaster, is a prime example of a significant toxic hotspot. The toxic gas leaked from the understaffed Union Carbide plant killed up to 20,000 people and left 120,000 others chronically ill. Bhopal continues to face pollution problems from the abandoned factory today.

### Air Pollution Hotspots

Air pollution hotspots are areas where air pollution emissions expose individuals to increased negative health effects. Hotspots denote areas in which a population's exposure to pollution and estimated health risks are high. Air pollution hotspots are particularly common in highly populated, urban areas, where there may a combination of stationary sources (e.g. industrial facilities) and mobile sources (e.g. cars and trucks) of pollution. Emissions from these sources can cause respiratory disease, childhood asthma, cancer, and other health problems. Fine particulate matter such as diesel soot, which contributes to more than 3.2 million premature deaths around the world each year, is a significant problem. It is very small and can lodge itself within the lungs and enter the bloodstream. Diesel soot is concentrated in densely populated areas, and one in six people in the U.S. live near a diesel pollution hot spot.

While air pollution hotspots affect a variety of populations, some groups are more likely to be located in hotspots. Previous studies have shown disparities in exposure to pollution by race and/or income (cite one of the early readings from our syllabus, e.g. Mohai & Pellow, or Saha). Hazardous land uses (toxic storage and disposal facilities, manufacturing facilities, major roadways) tend to be located where property values and income levels are low. Low socioeconomic status can be a proxy for other kinds of social vulnerability, including race, a lack of ability to influence regulatory permitting and a lack of ability to move to neighborhoods with less environmental pollution. These communities bear a disproportionate burden of environmental pollution and are more likely to face health risks such as cancer or asthma.

Studies show that patterns in race and income disparities not only indicate a higher exposure to pollution but also higher risk of adverse health outcomes. Communities characterized by low socioeconomic status and racial minorities can be more vulnerable to cumulative adverse health impacts resulting from elevated exposure to pollutants than more privileged communities. Blacks and Latinos generally face more pollution than whites and Asians, and low-income communities bear a higher burden of risk than affluent ones. Racial discrepancies are particularly distinct in suburban areas of the South and metropolitan areas of the West. Residents in public housing, who are generally low-income with poor access to health care and cannot move to healthier neighborhoods, are highly affected by nearby refineries and chemical plants.

Community groups and academic researchers have argued the unequal distribution of pollution on the poor and communities of color is an “environmental justice”.

Policy makers and researchers concerned with improving environmental justice for communities situated next to major sources of air pollution have developed a number of regulatory tools to identify air pollution hotspots. The EPA, for example, utilizes their Risk-Screening Environmental Indictors (RSEI) model to identify hotspots from a score of 3 to 15, with higher scores indicating closer proximity to hazards. Individual states have also taken steps to improve identification and surveillance. California’s AB 2588 Air Toxics “Hot Spots” Program, enacted in 1987, seeks to collect emission data, determine health risks, and notify local residents of major risks. By identifying hotspots regulators hope these tools will help them reduce pollution and inform nearby populations through the health risk assessments of individual pollutants and facilities that are summed in each zone to develop a total lifetime cancer risk. Air pollution hot spots are also at issue in pollution-trading programs, such as cap-and-trade systems designed to control pollution. These programs can potentially exacerbate effects from air pollution hotspots if the differences in chemical hazards are ignored. These programs also cause pollution to be mitigated towards where credit-buying firms are located. Factories can purchase emissions reduction credits from other firms, which leads to concentrated areas of pollution, since facilities that sell their credits are “exporting" their pollution to firms more likely to buy credits. However, some studies have noted that these claims have not materialized. Evan Ringquist, a professor at Indiana University of Public and Environmental Affairs, states that there is little empirical evidence to suggest the emergence of hotspots.

## West Oakland, California

Located in the East San Francisco Bay, the neighborhood of West Oakland is home to mainly low-income African American and Latino residents who are exposed to a disproportionate amount of airborne toxins as compared to the rest of the surrounding Alameda County. West Oakland's close proximity to highways and the Port of Oakland leave residents highly exposed to pollutants caused by moving and stationary sources of diesel pollution, thus leaving them at higher risk for health complications such as asthma and even shorter life expectancy than surrounding neighborhoods averages.

High emissions of toxic chemicals and airborne particulate matter in West Oakland that cause health issues are due to diesel fuels used for transportation in the Port of Oakland and surrounding highways. Traffic and transportation related air pollutants include carbon monoxide, nitrogen dioxide, black carbon, and diesel particulate matter. Residents are more exposed to harmful pollutants compared to other areas of the Bay Area and Oakland and therefore more at risk for harmful health effects. Compared to the State of California, West Oakland produces 90 times more diesel emission particulates per square mile per day. These pollutants have detrimental health effects such as asthma and reduced life expectancy while putting children at higher susceptibility for health complications.

Inequitable economic, residential, and environmental conditions in this low-income community of color leave residents of West Oakland with poor and inequitable health outcomes. African-American and Latino children of 10–18 years in West Oakland are more susceptible to onset lung defects such as asthma. According to Alameda County Vital Statistics, an African American child born in West Oakland is expected to live 14 fewer years than a white child born in the more wealthy Oakland Hills. Children 5 and under in west Oakland visit the emergency room for asthma three times more often than children in the county as a whole.

There are multiple efforts and strategies to spur legislation for equitable environmental conditions in low-income communities. There are many environmental justice groups and organizations in the Bay Area that encourage community participation in pursuing environmental justice. For example, data is collected by a Community-based participatory research (CBPR) and collaborated with West Oakland Environmental Indicators Project (WOEIP) in order to find effective and accurate findings to prove injustice and eventually spur reform in environmental policy. These research efforts can be used to document and communicate trends in air quality in West Oakland to policy makers. Effectiveness of efforts by these groups are multiplied by and increasing availability of environmental poverty lawyers who empower legislation in the legal system.

## Richmond, California

Richmond, located in the San Francisco Bay Area, is an evolving, multi-cultural community that has transformed itself from an over-polluted industrial town to a pioneer in an environmental justice movement. The city has been host to numerous oil refineries, including the Chevron Corporation refinery, which opened in 1901 under the ownership of Pacific Coast Oil. The Chevron Refinery is a leading source of air quality violations in the state of California. Richmond residents are also exposed to pollution from the Santa Fe train line and the presence of heavy traffic and diesel trucks along the Richmond Parkway. However, residents are most concerned with air pollution health impacts from the Chevron Refinery. In 1999, Richmond measure significantly higher on Air Quality Indices (AQI) (an indicator of how polluted is air is) compared to national level. Air pollution emission from the Chevron refinery includes benzene, ethyl benzene, toluene, xylene, nitrous dioxide, and sulfur dioxide, which are known to cause elevated cancer risks and respiratory illness. Rates of child and adult asthma are especially elevated among Richmond residents.

Richmond residents have struggled to improve local air quality. The city has a significant non-white, low-income population. According to 2010 U.S. Census, of Richmond's 103,701 person populations, "one in six residents lives below the federal poverty level, and more than eight in 10 are people of color. In North Richmond, next to one of the nation's largest refineries, 97 percent of residents are non-white and nearly one in four live in poverty". Low-income communities have differential access to political power, and their collective political voice is often less able to contest decisions impacting industrial operations. The combination of poverty, poor access to clean air, and poor political power can result in inequality in which communities of color bear a disproportionate burden of pollution and, therefore, suffer from greater environmental health risks.

Because Richmond is an air pollution hotspot, Richmond residents have applied different strategies since the 1980s to try to improve local air quality. The first EJ movement in the area started in the late 1980s, when the activist tried to stop construction of a garbage incinerator near North Richmond. Sixteen years later, local citizen utilizes "Bucket Brigades" to document a handful of criteria air pollutants such as sulfur dioxide [SO2], carbon monoxide [CO], nitrous dioxide [NO2], and ozone [O3].This study involves citizens to actively collecting the samples of emissions from Chevron's refineries, especially during accidents, fires, leaks, and explosions.The "sniffers" alert the "samplers" to collect the air samples when they notice a problem.The "samplers" then contact the Coordinator to check the bucket and perform the paperwork before submitting the samples to the Laboratory, in which the results will be reported to CBE, an environmental justice organization. The "Bucket Brigades" did not only raise the awareness local citizens to fight against the air pollution in their area, but also their participation.

As the number of activists and participants grew in numbers, their position in the battle against environmental injustice was further fortified with the election Green-party mayor of Richmond, Gayle McLaughlin, as well as three new council members sympathetic to their cause in 2008. In July 2008, despite the council failure to halt the Chevron's plan to build more refineries in the area due to rising gasoline prices during that time, the council succeeded to acquire $61 million from the oil company for community programs.

Due to great forces from the local communities and fellow EJ activists in Richmond area, Chevron has been making progress to embrace cleaner environment. In 2005, local activists managed to convince Bay Area Air Quality Management District to tighten the air pollution regulations by increasing the frequency of fines of facility incidents. Since then, Chevron has been flaring 10 times less than before. On top of that, Chevron has invested $150 million for building gas turbine in order to reduce air emission, increase energy efficiency, as well as provide most electrical and steam power Chevron requires to operate.

## Wilmington, Los Angeles

Bonnoris noted, "The environmental justice movement posits that the distribution of environmental harms and benefits should be fairly apportioned among all communities". As Bonnoris argued, the burden of air pollution is disproportionally distributed among communities based on their racial, social or economic status. Disproportion distribution of air pollution among communities can be a violation of the Equal Protection Clause of the Constitution because it violates equal protection of residents' public health.

Los Angeles is known for the nation's worst air quality and its "sharp inequalities in environmental exposures". Wilmington, Los Angeles is a neighborhood located on the southern part of Los Angeles, California. 54,512 people live in Wilmington, the median household income is $40,627, about 86 percentage of them are Latino and only 5.1% of Wilmington residents 25 or older have a four-year degree.

Wilmington, most of its residents are ethnic minorities, is possible to bear more environmental burden than other communities in Los Angeles because it is located next to several sources of air pollution. For example, Wilmington has "the highest concentration of refineries in the State". Emissions from refineries in Wilmington include carbon dioxide, sulfur dioxide and benzene. Wilmington has higher concentration of diesel particulate matter due to emissions from diesel trucks from the ports of Los Angeles and Long Beach. The risks associated with diesel are often underestimated since existing epidemiological studies cannot isolate exposure to diesel PM. However, exposure to diesel particulate matter can cause "irritation to the eyes, nose, throat and lungs", asthma, "exhaust immunological effects", and cancer.

Several NGOs have worked to improve the accuracy of Wilmington air quality data and air quality in order to protect approximately 1400 children who live or visit schools or childcare facilities at Wilmington. The environmental group "Coalition For a Safe Environment" installed an air pollution monitoring devices on the residential buildings in Wilmington in order to prove that emissions from local oil refineries and diesel trucks to the ports pollute the air in Wilmington, disproportionately affecting Wilmington residents to suffer from health problems including lung diseases and respiratory diseases.

## Groundwater Contamination

The town of Hinkley, California, located in the Mojave Desert, had its groundwater contaminated with hexavalent chromium starting in 1952, resulting in a legal case against Pacific Gas & Electric (PG&E) and a multimillion-dollar settlement in 1996. The legal case was dramatized in the film *Erin Brockovich*, released in 2000.

PG&E operates a compressor station in Hinkley for natural gas transmission pipelines. The natural gas has to be re-compressed approximately every 350 miles (560 km), and the station uses large cooling towers to cool the gas after it has been compressed. Between 1952 and 1966, the water used in these cooling towers contained hexavalent chromium – now recognized as a carcinogen – to prevent rust in the machinery. The water was stored between uses in unlined ponds, which allowed it to percolate into the groundwater. This severely contaminated the groundwater, affecting soil and contaminating water wells near the compressor station, with a plume approximately 2 miles (3.2 km) long and nearly 1 mile (1.6 km) wide.

## Radioactive Contamination

### Pacific Proving Grounds

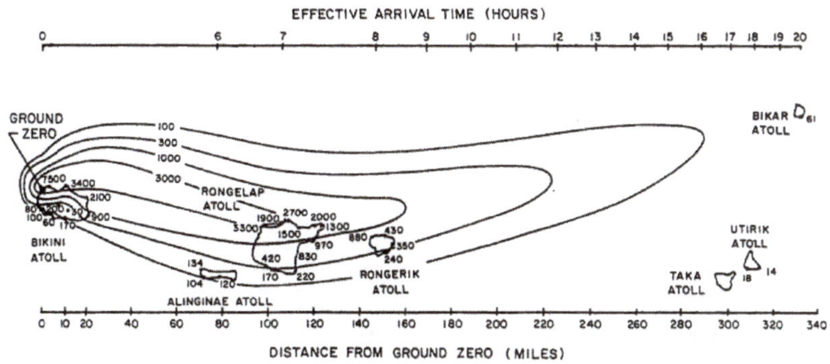


The Castle Bravo test of 1954 spread nuclear fallout across the Marshall Islands, parts of which were still inhabited.

The Pacific Proving Grounds was the name used to describe a number of sites in the Marshall Islands and a few other sites in the Pacific Ocean, used by the United States to conduct nuclear testing at various times between 1946 and 1962. In July 1947, after the first atomic weapons testing at Bikini Atoll, the United States entered into an agreement with the United Nations to govern the Trust Territory of the Pacific Islands as a strategic trusteeship territory. The Trust Territory is composed of 2,000 islands spread over 3,000,000 square miles (7,800,000 km²) of the North Pacific Ocean. On July 23, 1947, the United States Atomic Energy Commission announced the establishment of the Pacific Proving Grounds.

105 atmospheric (i.e., not underground) nuclear tests were conducted there, many of which were of extremely high yield. While the Marshall Islands testing composed 14% of all U.S. tests, it composed nearly 80% of the total yields of those detonated by the U.S., with an estimated total yield of around 210 megatons, with the largest being the 15 Mt Castle Bravo shot of 1954 which spread considerable nuclear fallout on many of the islands, including several which were inhabited, and some that had not been evacuated.

Many of the islands which were part of the Pacific Proving Grounds continue to be contaminated by nuclear fallout, and many of those who were living on the islands at the time of testing have suffered from an increased incidence of various health problems. Through the Radiation Exposure Compensation Act of 1990, at least $759 million has been paid to Marshall Islanders as compensation for their exposure to U.S. nuclear testing. Following the Castle Bravo accident, $15.3 million was paid to Japan.

### Nevada Test Site

The Nevada Test Site (NTS), is a United States Department of Energy reservation located in southeastern Nye County, Nevada, about 65 miles (105 km) northwest of the city of Las Vegas. Formerly known as the Nevada Proving Grounds, the site was established on 11 January 1951 for the testing of nuclear devices, covering approximately 1,360 square miles (3,500 km²) of desert and mountainous terrain. Nuclear testing at the Nevada Test Site began with a 1-kiloton-of-TNT (4.2 TJ) bomb dropped on Frenchman Flat on 27 January 1951. Many of the iconic images of the nuclear era come from the NTS.

Mushroom cloud from the Nevada Test Site seen from downtown Las Vegas.

During the 1950s, the mushroom clouds from atmospheric tests could be seen for almost 100 mi (160 km). The city of Las Vegas experienced noticeable seismic effects, and the distant mushroom clouds, which could be seen from the downtown hotels, became tourist attractions. St. George, Utah, received the brunt of the fallout of above-ground nuclear testing in the Yucca Flats/Nevada Test Site. Winds routinely carried the fallout of these tests directly through St. George and southern Utah. Marked increases in cancers, such as leukemia, lymphoma, thyroid cancer, breast cancer, melanoma, bone cancer, brain tumors, and gastrointestinal tract cancers, were reported from the mid-1950s through 1980.

From 1986 through 1994, two years after the United States put a hold on full-scale nuclear weapons testing, 536 anti-nuclear protests were held at the Nevada Test Site involving 37,488 participants and 15,740 arrests, according to government records. Those arrested included the astronomer Carl Sagan and the actors Kris Kristofferson, Martin Sheen, and Robert Blake.

The Nevada Test Site contains 28 areas, 1,100 buildings, 400 miles (640 km) of paved roads, 300 miles of unpaved roads, ten heliports, and two airstrips. The most-recent test was a sub-critical test of the properties of plutonium, conducted underground on December 7, 2012.

## Semipalatinsk Test Site

The Semipalatinsk Test Site, also known as "The Polygon", was the primary testing venue for the Soviet Union's nuclear weapons. It is located on the steppe in northeast Kazakhstan (then the Kazakh SSR), south of the valley of the Irtysh River. The scientific buildings for the test site were located around 150 km west of the town of Semipalatinsk (later renamed Semey), near the border of East Kazakhstan Province and Pavlodar Province with most of the nuclear tests taking place at various sites further to the west and south, some as far as into Karagandy Province.

The Soviet Union conducted 456 nuclear tests at Semipalatinsk from 1949 until 1989 with little regard for their effect on the local people or environment. The full impact of radiation exposure was hidden for many years by Soviet authorities and has only come to light since the test site closed in 1991.

From 1996 to 2012, a secret joint operation of Kazakh, Russian, and American nuclear scientists and engineers secured the waste plutonium in the tunnels of the mountains.

## Environmental Degradation

Eighty-plus years after the abandonment of Wallaroo Mines (Kadina, South Australia), mosses remain the only vegetation at some spots of the site's grounds

Environmental degradation is the deterioration of the environment through depletion of resources such as air, water and soil; the destruction of ecosystems; habitat destruction; the extinction of wildlife; and pollution. It is defined as any change or disturbance to the environment perceived to be deleterious or undesirable. As indicated by the I=PAT equation, environmental impact (I) or degradation is caused by the combination of an already very large and increasing human population (P), continually increasing economic growth or per capita affluence (A), and the application of resource depleting and polluting technology (T).

Environmental degradation is one of the ten threats officially cautioned by the High-level Panel on Threats, Challenges and Change of the United Nations. The United Nations International Strategy for Disaster Reduction defines environmental degradation as "The reduction of the capacity of the environment to meet social and ecological objectives, and needs". Environmental degradation is of many types. When natural habitats are destroyed or natural resources are depleted, the environment is degraded. Efforts to counteract this problem include environmental protection and environmental resources management.

### Water Degradation

One major component of environmental degradation is the depletion of the resource of fresh water on Earth. Approximately only 2.5% of all of the water on Earth is fresh water, with the rest being salt water. 69% of the fresh water is frozen in ice caps located on Antarctica and Greenland, so only 30% of the 2.5% of fresh water is available for consumption. Fresh water is an exceptionally important resource, since life on Earth is ultimately dependent on it. Water transports nutrients and chemicals within the biosphere to all forms of life, sustains both plants and animals, and moulds the surface of the Earth with transportation and deposition of materials.

The current top three uses of fresh water account for 95% of its consumption; approximately 85% is used for irrigation of farmland, golf courses, and parks, 6% is used for domestic purposes such as indoor bathing uses and outdoor garden and lawn use, and 4% is used for industrial purposes such as processing, washing, and cooling in manufacturing centers. It is estimated that one in three people over the entire globe are already facing water shortages, almost one-fifth of the world's population live in areas of physical water scarcity, and almost one quarter of the world's population live in a developing country that lacks the necessary infrastructure to use water from available rivers and aquifers. Water scarcity is an increasing problem due to many foreseen issues in the future, including population growth, increased urbanization, higher standards of living, and climate change.

## Climate Change and Temperature

Climate change affects the Earth's water supply in a large number of ways. It is predicted that the mean global temperature will rise in the coming years due to a number of forces affecting the climate, the amount of atmospheric $CO_2$ will rise, and both of these will influence water resources; evaporation depends strongly on temperature and moisture availability, which can ultimately affect the amount of water available to replenish groundwater supplies.

Transpiration from plants can be affected by a rise in atmospheric $CO_2$, which can decrease their use of water, but can also raise their use of water from possible increases of leaf area. Temperature increase can decrease the length of the snow season in the winter and increase the intensity of snowmelt in warmer seasons, leading to peak runoff of snowmelt earlier in the season, affecting soil moisture, flood and drought risks, and storage capacities depending on the area.

Warmer winter temperatures cause a decrease in snowpack, which can result in diminished water resources during summer. This is especially important at mid-latitudes and in mountain regions that depend on glacial runoff to replenish their river systems and groundwater supplies, making these areas increasingly vulnerable to water shortages over time; an increase in temperature will initially result in a rapid rise in water melting from glaciers in the summer, followed by a retreat in glaciers and a decrease in the melt and consequently the water supply every year as the size of these glaciers get smaller and smaller.

Thermal expansion of water and increased melting of oceanic glaciers from an increase in temperature gives way to a rise in sea level, which can affect the fresh water supply of coastal areas as well; as river mouths and deltas with higher salinity get pushed further inland, an intrusion of saltwater results in an increase of salinity in reservoirs and aquifers. Sea-level rise may also consequently be caused by a depletion of groundwater, as climate change can affect the hydrologic cycle in a number of ways. Uneven distributions of increased temperatures and increased precipitation around the globe results in water surpluses and deficits, but a global decrease in groundwater suggests a rise in sea level, even after meltwater and thermal expansion were accounted for, which can provide a positive feedback to the problems sea-level rise causes to fresh-water supply.

A rise in air temperature results in a rise in water temperature, which is also very significant in water degradation, as the water would become more susceptible to bacterial growth. An increase in water temperature can also affect ecosystems greatly because of a species' sensitivity to temperature, and also by inducing changes in a body of water's self-purification system from decreased amounts of dissolved oxygen in the water due to rises in temperature.

## Climate Change and Precipitation

A rise in global temperatures is also predicted to correlate with an increase in global precipitation, but because of increased runoff, floods, increased rates of soil erosion, and mass movement of land, a decline in water quality is probable, while water will carry more nutrients, it will also carry more contaminants. While most of the attention about climate change is directed towards global warming and greenhouse effect, some of the most severe effects of climate change are likely to be from changes in precipitation, evapotranspiration, runoff, and soil moisture. It is generally expected that, on average, global precipitation will increase, with some areas receiving increases and some decreases.

Climate models show that while some regions should expect an increase in precipitation, such as in the tropics and higher latitudes, other areas are expected to see a decrease, such as in the subtropics; this will ultimately cause a latitudinal variation in water distribution. The areas receiving more precipitation are also expected to receive this increase during their winter and actually become drier during their summer, creating even more of a variation of precipitation distribution. Naturally, the distribution of precipitation across the planet is very uneven, causing constant variations in water availability in respective locations.

Changes in precipitation affect the timing and magnitude of floods and droughts, shift runoff processes, and alter groundwater recharge rates. Vegetation patterns and growth rates will be directly affected by shifts in precipitation amount and distribution, which will in turn affect agriculture as well as natural ecosystems. Decreased precipitation will deprive areas of water, causing water tables to fall and reservoirs and wetlands, rivers, and lakes to empty, and possibly an increase in evaporation and evapotranspiration, depending on the accompanied rise in temperature. Groundwater reserves will be depleted, and the remaining water has a greater chance of being of poor quality from saline or contaminants on the land surface.

## Population Growth

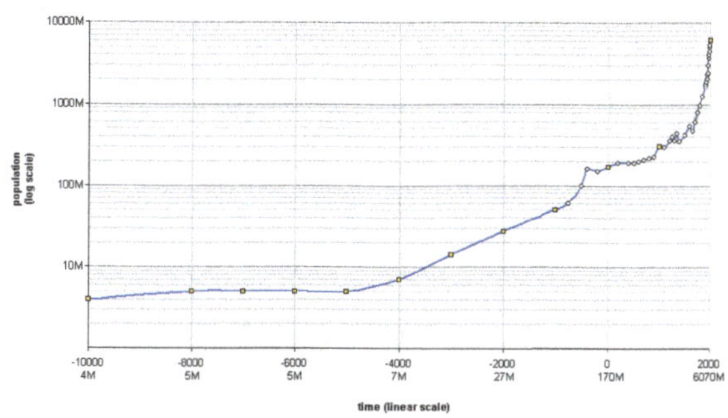


World population growth in a lin-log scale.

The human population on Earth is expanding rapidly which goes hand in hand with the degradation of the environment at large measures. Humanity's appetite for needs is disarranging the environment's natural equilibrium. Production industries are venting smoke and discharging chemicals that are polluting water resources. The smoke that is emitted into the atmosphere holds detrimental gases such as carbon monoxide and sulfur dioxide. The high levels of pollution in the atmosphere form layers that are eventually absorbed into the atmosphere. Organic compounds such as chlorofluorocarbons (CFC's) have generated an unwanted opening in the ozone layer, which emits higher levels of ultraviolet radiation putting the globe at large threat.

The available fresh water being affected by the climate is also being stretched across an ever-increasing global population. It is estimated that almost a quarter of the global population is living in an area that is using more than 20% of their renewable water supply; water use will rise with population while the water supply is also being aggravated by decreases in streamflow and groundwater caused by climate change. Even though some areas may see an increase in freshwater supply from an uneven distribution of precipitation increase, an increased use of water supply is expected.

An increased population means increased withdrawals from the water supply for domestic, agricultural, and industrial uses, the largest of these being agriculture, believed to be the major non-climate driver of environmental change and water deterioration. The next 50 years will likely be the last period of rapid agricultural expansion, but the larger and wealthier population over this time will demand more agriculture.

Population increase over the last two decades, at least in the United States, has also been accompanied by a shift to an increase in urban areas from rural areas, which concentrates the demand for water into certain areas, and puts stress on the fresh water supply from industrial and human contaminants. Urbanization causes overcrowding and increasingly unsanitary living conditions, especially in developing countries, which in turn exposes an increasingly number of people to disease. About 79% of the world's population is in developing countries, which lack access to sanitary water and sewer systems, giving rises to disease and deaths from contaminated water and increased numbers of disease-carrying insects.

## Agriculture

Water pollution due to dairy farming in the Wairarapa in New Zealand

Agriculture is dependent on available soil moisture, which is directly affected by climate dynamics, with precipitation being the input in this system and various processes being the output, such as evapotranspiration, surface runoff, drainage, and percolation into groundwater. Changes in climate, especially the changes in precipitation and evapotranspiration predicted by climate models, will directly affect soil moisture, surface runoff, and groundwater recharge.

In areas with decreasing precipitation as predicted by the climate models, soil moisture may be substantially reduced. With this in mind, agriculture in most areas needs irrigation already, which depletes fresh water supplies both by the physical use of the water and the degradation agriculture causes to the water. Irrigation increases salt and nutrient content in areas that would not normally be affected, and damages streams and rivers from damming and removal of water. Fertilizer enters both human and livestock waste streams that eventually enter groundwater, while nitrogen, phosphorus, and other chemicals from fertilizer can acidify both soils and water. Certain agricultural demands may increase more than others with an increasingly wealthier global population, and meat is one commodity expected to double global food demand by 2050, which directly affects the global supply of fresh water. Cows need water to drink, more if the temperature is high and humidity is low, and more if the production system the cow is in is extensive, since finding food takes more effort. Water is needed in processing of the meat, and also in the production of feed for the livestock. Manure can contaminate bodies of freshwater, and slaughterhouses, depending on how well they are managed, contribute waste such as blood, fat, hair, and other bodily contents to supplies of fresh water.

The transfer of water from agricultural to urban and suburban use raises concerns about agricultural sustainability, rural socioeconomic decline, food security, an increased carbon footprint from imported food, and decreased foreign trade balance. The depletion of fresh water, as applied to more specific and populated areas, increases fresh water scarcity among the population and also makes populations susceptible to economic, social, and political conflict in a number of ways; rising sea levels forces migration from coastal areas to other areas farther inland, pushing populations closer together breaching borders and other geographical patterns, and agricultural surpluses and deficits from the availability of water induce trade problems and economies of certain areas. Climate change is an important cause of involuntary migration and forced displacement According to the Food and Agriculture Organization of the United Nations, global greenhouse gas emissions from animal agriculture exceeds that of transportation.

## Water Management

The issue of the depletion of fresh water can be met by increased efforts in water management. While water management systems are often flexible, adaptation to new hydrologic conditions may be very costly. Preventative approaches are necessary to avoid high costs of inefficiency and the need for rehabilitation of water supplies, and innovations to decrease overall demand may be important in planning water sustainability.

Water supply systems, as they exist now, were based on the assumptions of the current climate, and built to accommodate existing river flows and flood frequencies. Reservoirs are operated based on past hydrologic records, and irrigation systems on historical temperature, water availability, and crop water requirements; these may not be a reliable guide to the future. Re-examining engineering designs, operations, optimizations, and planning, as well as re-evaluating legal, technical, and economic approaches to manage water resources are very important for the future of water management in response to water degradation. Another approach is water privatization; despite its economic and cultural effects, service quality and overall quality of the water can be more easily controlled and distributed. Rationality and sustainability is appropriate, and requires limits to overexploitation and pollution, and efforts in conservation.

## References


- Miller K. A.; Siscovick D. S.; Sheppard L.; Shepherd K.; Sullivan J. H.; Anderson G. L.; Kaufman J. D. (2007). "Long-term exposure to air pollution and incidence of cardiovascular events in women.". The New England Journal of Medicine
- State Coalition for Remediation of Drycleaners (2011). "A Citizen's Guide to Drycleaner Cleanup" (PDF). Retrieved 20 March 2017
- Goel, P.K. (2006). Water Pollution - Causes, Effects and Control. New Delhi: New Age International. p. 179. ISBN 978-81-224-1839-2
- Health; Occupational; Thoracic Society, American (1996). "[Comparative Study Review]". American Journal of Respiratory and Critical Care Medicine. 153 (1): 3–50
- "Texas fracking site that spilled 42,000 gallons of fluid into residential area hopes to reopen". RT International. Retrieved 2016-05-07
- Kennish, Michael J. (1992). Ecology of Estuaries: Anthropogenic Effects. Marine Science Series. Boca Raton, FL: CRC Press. pp. 415–17. ISBN 978-0-8493-8041-9
- J. Sunyer (2001). "Urban air pollution and Chronic Obstructive Pulmonary disease: a review". European Respiratory Journal. 17 (5): 1024–1033
- "Lack of data on fracking spills leaves researchers in the dark on water contamination". StateImpact Pennsylvania. Retrieved 2016-05-09
- Laws, Edward A. (2000). Aquatic Pollution: An Introductory Text. New York: John Wiley and Sons. p. 430. ISBN 978-0-471-34875-7
- J. C. Fensterstock , J. A. Kurtzweg & G. Ozolins (1971): Reduction of Air Pollution Potential through Environmental Planning, Journal of the Air Pollution Control Association, 21:7, 395-399
- World Health Organization (WHO) (2006). "Protecting Groundwater for Health - Understanding the drinking-water catchment" (PDF). Retrieved 20 March 2017
- For example, see Baird, Rodger B.; Clesceri, Leonore S.; Eaton, Andrew D.; et al., eds. (2012). Standard Methods for the Examination of Water and Wastewater (22nd ed.). Washington, DC: American Public Health Association. ISBN 978-0875530130
- Mohai, P; Lantz, PM; Morenoff, J; House, JS; Mero, RP (2009). "Racial and Socioeocnomic Disparities in Residential Proximity". American Journal of Public Health. 99 (3): S649–S656
- Flagstaff Dark Skies Coalition. "Lamp Spectrum and Light Pollution". Lamp Spectrum and Light Pollution. Retrieved 10 April 2016
- Goel, P.K. (2006). Water Pollution - Causes, Effects and Control. New Delhi: New Age International. ISBN 978-81-224-1839-2
- J.M. Field, Effect of personal and situational variables upon noise annoyance in residential areas, Journal of the Acoustical Society of America, 93: 2753-2763 (1993)
- United Nations Environment Programme (UNEP) (2015). "Good Practices for Regulating Wastewater Treatment" (PDF). Retrieved 19 March 2017
- Kennish, Michael J. (1992). Ecology of Estuaries: Anthropogenic Effects. Marine Science Series. Boca Raton, FL: CRC Press. ISBN 978-0-8493-8041-9
- Di Luzio, Frank C. (January 1967). "Water Pollution Control: An American Must". Journal (Water Pollution Control Federation). Water Environment Federation. 39 (1): 1–7. JSTOR 25035710

3

# Devices and Technologies for Air and Water Pollution Control

It is very important to control pollution to sustain the life of our planet. Some of the devices and technologies introduced for air and water pollution control are dust collector, scrubber, sewage treatment and vapor recovery. The chapter serves as a source to understand the devices and technologies used for controlling pollution.

## Dust Collector

Baghouse Dust Collector for Asphalt Plants

A dust collector is a system used to enhance the quality of air released from industrial and commercial processes by collecting dust and other impurities from air or gas. Designed to handle high-volume dust loads, a dust collector system consists of a blower, dust filter, a filter-cleaning system, and a dust receptacle or dust removal system. It is distinguished from air cleaners, which use disposable filters to remove dust.

### Uses

Dust collectors are used in many processes to either recover valuable granular solid or powder from process streams, or to remove granular solid pollutants from exhaust gases prior to venting to the atmosphere. Dust collection is an online process for collecting any process-generated dust from the source point on a continuous basis. Dust collectors may be of single unit construction, or a collection of devices used to separate particulate matter from the process air. They are often used as an air pollution control device to maintain or improve air quality.

Mist collectors remove particulate matter in the form of fine liquid droplets from the air. They are often used for the collection of metal working fluids, and coolant or oil mists. Mist collectors are often used to improve or maintain the quality of air in the workplace environment.

Fume and smoke collectors are used to remove sub-micrometer-size particulates from the air. They effectively reduce or eliminate particulate matter and gas streams from many industrial processes such as welding, rubber and plastic processing, high speed machining with coolants, tempering, and quenching.

## Types of Dust Collectors

Five main types of industrial dust collectors are:

- Inertial separators
- Fabric filters
- Wet scrubbers
- Unit collectors
- Electrostatic precipitators

## Inertial Separators

Inertial separators separate dust from gas streams using a combination of forces, such as centrifugal, gravitational, and inertial. These forces move the dust to an area where the forces exerted by the gas stream are minimal. The separated dust is moved by gravity into a hopper, where it is temporarily stored.

The three primary types of inertial separators are:

- Settling chambers
- Baffle chambers
- Centrifugal collectors

Neither settling chambers nor baffle chambers are commonly used in the minerals processing industry. However, their principles of operation are often incorporated into the design of more efficient dust collectors.

## Settling Chamber

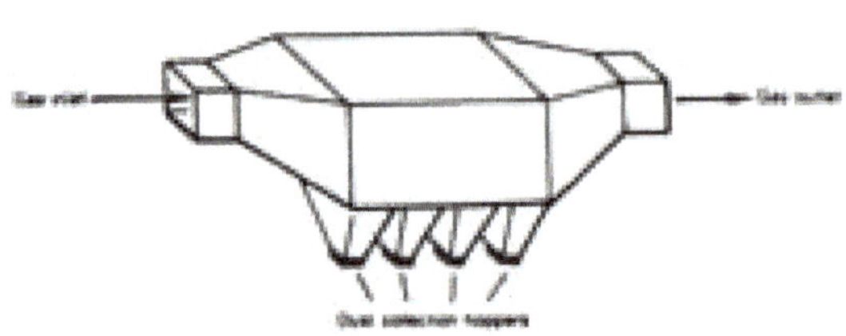

Settling Chamber

A settling chamber consists of a large box installed in the ductwork. The increase of cross section area at the chamber reduces the speed of the dust-filled airstream and heavier particles settle out. Settling chambers are simple in design and can be manufactured from almost any material. However, they are seldom used as primary dust collectors because of their large space requirements and low efficiency. A practical use is as precleaners for more efficient collect.

Advantages: 1. simple construction and low cost. 2. maintaince free. 3. collection of particle without need of water.

Disadvantages: 1. low efficiency. 2. large space required.

## Baffle Chamber

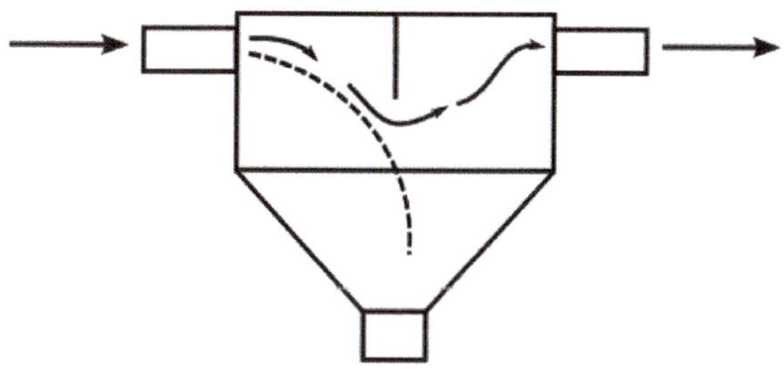

Baffle Chamber

Diagram of a Baffle Chamber

Baffle chambers use a fixed baffle plate that causes the conveying gas stream to make a sudden change of direction. Large-diameter particles do not follow the gas stream but continue into a dead air space and settle. Baffle chambers are used as precleaners

## Centrifugal Collectors

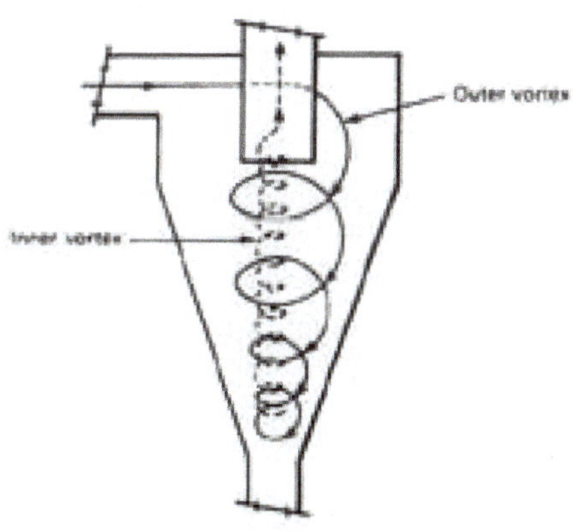


Cyclone

Centrifugal collectors use cyclonic action to separate dust particles from the gas stream. In a typical cyclone, the dust gas stream enters at an angle and is spun rapidly. The centrifugal force created by the circular flow throws the dust particles toward the wall of the cyclone. After striking the wall, these particles fall into a hopper located underneath.

The most common types of centrifugal, or inertial, collectors in use today are:-

## Single-cyclone Separators

They create a dual vortex to separate coarse from fine dust. The main vortex spirals downward and carries most of the coarser dust particles. The inner vortex, created near the bottom of the cyclone, spirals upward and carries finer dust particles.

## Multiple-cyclone Separators

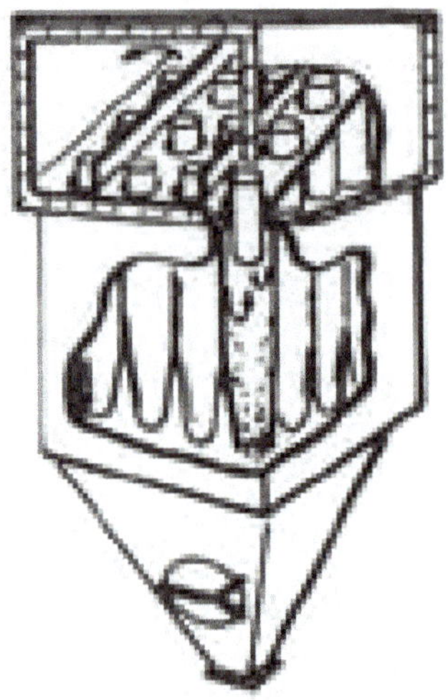

Multiclone

Multiple-cyclone separators consist of a number of small-diameter cyclones, operating in parallel and having a common gas inlet and outlet, as shown in the figure, and operate on the same principle as single cyclone separators—creating an outer downward vortex and an ascending inner vortex.

Multiple-cyclone separators remove more dust than single cyclone separators because the individual cyclones have a greater length and smaller diameter. The longer length provides longer residence time while the smaller diameter creates greater centrifugal force. These two factors result in better separation of dust particulates. The pressure drop of multiple-cyclone separators collectors is higher than that of single-cyclone separators, requiring more energy to clean the same amount of air. A single-chamber cyclone separator of the same volume is more economical, but doesn't remove as much dust.

Cyclone separators are found in all types of power and industrial applications, including pulp and paper plants, cement plants, steel mills, petroleum coke plants, metallurgical plants, saw mills and other kinds of facilities that process dust.

## Secondary-air-flow Separators

This type of cyclone uses a secondary air flow, injected into the cyclone to accomplish several things. The secondary air flow increases the speed of the cyclonic action making the separator more efficient; it intercepts the particulate before it reaches the interior walls of the unit; and it forces the separated particulate toward the collection area. The secondary air flow protects the separator from particulate abrasion and allows the separator to be installed horizontally because gravity is not depended upon to move the separated particulate downward.

## Fabric Filters

Commonly known as baghouses, fabric collectors use filtration to separate dust particulates from dusty gases. They are one of the most efficient and cost effective types of dust collectors available and can achieve a collection efficiency of more than 99% for very fine particulates.

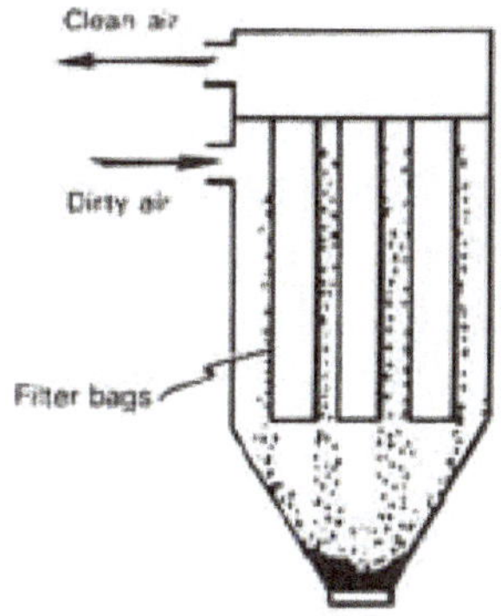


Baghouse

Dust-laden gases enter the baghouse and pass through fabric bags that act as filters. The bags can be of woven or felted cotton, synthetic, or glass-fiber material in either a tube or envelope shape.

## Pre-coating

To ensure the filter bags have a long usage life they are commonly coated with a filter enhancer (pre-coat). The use of chemically inert limestone (calcium carbonate) is most common as it maximises efficiency of dust collection (including fly ash) via formation of what is called a dustcake or coating on the surface of the filter media. This not only traps fine particulates but also provides protection for the bag itself from moisture, and oily or sticky particulates which can bind the filter media. Without a pre-coat the filter bag allows fine particulates to bleed through the bag filter system, especially during start-up, as the bag can only do part of the filtration leaving the finer parts to the filter enhancer dustcake.

## Parts

Fabric filters generally have the following parts:

1. Clean plenum
2. Dusty plenum
3. Bag, cage, venturi assembly
4. Tubeplate
5. RAV/SCREW
6. Compressed air header

7. Blow pipe
8. Housing and hopper

## Types of Bag Cleaning

Baghouses are characterized by their cleaning method.

### Shaking

A rod connecting to the bag is powered by a motor. This provides motion to remove caked-on particles. The speed and motion of the shaking depends on the design of the bag and composition of the particulate matter. Generally shaking is horizontal. The top of the bag is closed and the bottom is open. When shaken, the dust collected on the inside of the bag is freed. During the cleaning process, no dirty gas flows through a bag while the bag is being cleaned. This redirection of air flow illustrates why baghouses must be compartmentalized.

### Reverse Air

Air flow gives the bag structure. Dirty air flows through the bag from the inside, allowing dust to collect on the interior surface. During cleaning, gas flow is restricted from a specific compartment. Without the flowing air, the bags relax. The cylindrical bag contains rings that prevent it from completely collapsing under the pressure of the air. A fan blows clean air in the reverse direction. The relaxation and reverse air flow cause the dust cake to crumble and release into the hopper. Upon the completion of the cleaning process, dirty air flow continues and the bag regains its shape.

### Pulse Jet

This type of baghouse cleaning (also known as pressure-jet cleaning) is the most common. A high pressure blast of air is used to remove dust from the bag. The blast enters the top of the bag tube, temporarily ceasing the flow of dirty air. The shock of air causes a wave of expansion to travel down the fabric. The flexing of the bag shatters and discharges the dust cake. The air burst is about 0.1 second and it takes about 0.5 seconds for the shock wave to travel down the length of the bag. Due to its rapid release, the blast of air does not interfere with contaminated gas flow. Therefore, pulse-jet baghouses can operate continuously and are not usually compartmentalized. The blast of compressed air must be powerful enough to ensure that the shock wave will travel the entire length of the bag and fracture the dust cake.

### Sonic

The least common type of cleaning method is sonic. Shaking is achieved by sonic vibration. A sound generator produces a low frequency sound that causes the bags to vibrate. Sonic cleaning is commonly combined with another method of cleaning to ensure thorough cleaning.

## Cartridge Collectors

Cartridge collectors use perforated metal cartridges that contain a pleated, nonwoven filtering media, as opposed to woven or felt bags used in baghouses. The pleated design allows for a greater total filtering surface area than in a conventional bag of the same diameter, The greater filtering area results in a reduced air to media ratio, pressure drop, and overall collector size.

Cartridge collectors are available in single use or continuous duty designs. In single-use collectors, the dirty cartridges are changed and collected dirt is removed while the collector is off. In the continuous duty design, the cartridges are cleaned by the conventional pulse-jet cleaning system.

## Wet Scrubbers

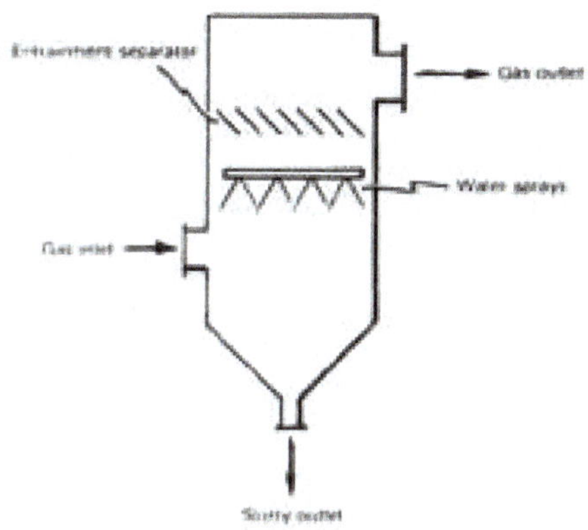

**Wet Scrubber**

Dust collectors that use liquid are known as wet scrubbers. In these systems, the scrubbing liquid (usually water) comes into contact with a gas stream containing dust particles. Greater contact of the gas and liquid streams yields higher dust removal efficiency.

There is a large variety of wet scrubbers; however, all have one of three basic configurations:

1. Gas-humidification - The gas-humidification process agglomerates fine particles, increasing the bulk, making collection easier.

2. Gas-liquid contact - This is one of the most important factors affecting collection efficiency. The particle and droplet come into contact by four primary mechanisms:

a) Inertial impaction - When water droplets placed in the path of a dust-laden gas stream, the stream separates and flows around them. Due to inertia, the larger dust particles will continue on in a straight path, hit the droplets, and become encapsulated.

b) Interception - Finer particles moving within a gas stream do not hit droplets directly but brush against and adhere to them.

c) Diffusion - When liquid droplets are scattered among dust particles, the particles are deposited on the droplet surfaces by Brownian movement, or diffusion. This is the principal mechanism in the collection of submicrometre dust particles.

d) Condensation nucleation - If a gas passing through a scrubber is cooled below the dew-point, condensation of moisture occurs on the dust particles. This increase in particle size makes collection easier.

3 Gas-liquid separation - Regardless of the contact mechanism used, as much liquid and dust as possible must be removed. Once contact is made, dust particulates and water droplets combine to form agglomerates. As the agglomerates grow larger, they settle into a collector.

The “cleaned” gases are normally passed through a mist eliminator (demister pads) to remove water droplets from the gas stream. The dirty water from the scrubber system is either cleaned and discharged or recycled to the scrubber. Dust is removed from the scrubber in a clarification unit or a drag chain tank. In both systems solid material settles on the bottom of the tank. A drag chain system removes the sludge and deposits in into a dumpster or stockpile.

## Types of Scrubbers

Spray-tower scrubber wet scrubbers may be categorized by pressure drop as follows:

- Low-energy scrubbers (0.5 to 2.5 inches water gauge - 124.4 to 621.9 Pa)
- Low- to medium-energy scrubbers (2.5 to 6 inches water gauge - 0.622 to 1.493 kPa)
- Medium- to high-energy scrubbers (6 to 15 inches water gauge - 1.493 to 3.731 kPa)
- High-energy scrubbers (greater than 15 inches water gauge - greater than 3.731 kPa)

Due to the large number of commercial scrubbers available, it is not possible to describe each individual type here.

## Low-energy Scrubbers

In the simple, gravity-spray-tower scrubber, liquid droplets formed by liquid atomized in spray nozzles fall through rising exhaust gases. Dirty water is drained at the bottom.

These scrubbers operated at pressure drops of 1 to 2 in. water gauge (¼ to ½ kPa) and are approximately 70% efficient on 10 μm particles. Their efficiency is poor below 10 μm. However, they are capable of treating relatively high dust concentrations without becoming plugged.

## Low- to Medium-energy Scrubbers

Wet cyclones use centrifugal force to spin the dust particles (similar to a cyclone), and throw the particulates upon the collector’s wetted walls. Water introduced from the top to wet the cyclone walls carries these particles away. The wetted walls also prevent dust reentrainment.

Pressure drops for these collectors range from 2 to 8 in. water (½ to 2 kPa), and the collection efficiency is good for 5 μm particles and above.

## high-energy Scrubbers co-current-flow Scrubber

Packed-bed scrubbers consist of beds of packing elements, such as coke, broken rock, rings, saddles, or other manufactured elements. The packing breaks down the liquid flow into a high-surface-area film so that the dusty gas streams passing through the bed achieve maximum contact with the liquid film and become deposited on the surfaces of the packing elements. These scrubbers have a good collection efficiency for respirable dust.

Three types of packed-bed scrubbers are-

- Cross-flow scrubbers
- Co-current flow scrubbers
- Counter-current flow scrubbers

Efficiency can be greatly increased by minimizing target size, i.e., using 0.003 in. (0.076 mm) diameter stainless steel wire and increasing gas velocity to more than 1,800 ft/min (9.14 m/s).

## High-energy Scrubbers

Venturi scrubbers consist of a venturi-shaped inlet and separator. The dust-laden gases venturi scrubber enter through the venturi and are accelerated to speeds between 12,000 and 36,000 ft/min (60.97-182.83 m/s). These high-gas velocities immediately atomize the coarse water spray, which is injected radially into the venturi throat, into fine droplets. High energy and extreme turbulence promote collision between water droplets and dust particulates in the throat. The agglomeration process between particle and droplet continues in the diverging section of the venturi. The large agglomerates formed in the venturi are then removed by an inertial separator.

Venturi scrubbers achieve very high collection efficiencies for respirable dust. Since efficiency of a venturi scrubber depends on pressure drop, some manufacturers supply a variable-throat venturi to maintain pressure drop with varying gas flows.

## Electrostatic Precipitators (ESP)

Electrostatic precipitators use electrostatic forces to separate dust particles from exhaust gases. A number of high-voltage, direct-current discharge electrodes are placed between grounded collecting electrodes. The contaminated gases flow through the passage formed by the discharge and collecting electrodes. Electrostatic precipitators operate on the same principle as home "Ionic" air purifiers.

The airborne particles receive a negative charge as they pass through the ionized field between the electrodes. These charged particles are then attracted to a grounded or positively charged electrode and adhere to it.

The collected material on the electrodes is removed by rapping or vibrating the collecting electrodes either continuously or at a predetermined interval. Cleaning a precipitator can usually be done without interrupting the airflow.

The four main components of all electrostatic precipitators are:

- Power supply unit, to provide high-voltage DC power
- Ionizing section, to impart a charge to particulates in the gas stream

- A means of removing the collected particulates
- A housing to enclose the precipitator zone

The following factors affect the efficiency of electrostatic precipitators:

- Larger collection-surface areas and lower gas-flow rates increase efficiency because of the increased time available for electrical activity to treat the dust particles.
- An increase in the dust-particle migration velocity to the collecting electrodes increases efficiency. The migration velocity can be increased by:
    - Decreasing the gas viscosity
    - Increasing the gas temperature
    - Increasing the voltage field

## Types of Precipitators

There are two main types of precipitators:

- High-voltage, single-stage - Single-stage precipitators combine an ionization and a collection step. They are commonly referred to as Cottrell precipitators.
- Low-voltage, two-stage - Two-stage precipitators use a similar principle; however, the ionizing section is followed by collection plates.

Described below is the high-voltage, single-stage precipitator, which is widely used in minerals processing operations. The low-voltage, two-stage precipitator is generally used for filtration in air-conditioning systems.

## Plate Precipitators

The majority of electrostatic precipitators installed are the plate type. Particles are collected on flat, parallel surfaces that are 8 to 12 in. (20 to 30 cm) apart, with a series of discharge electrodes spaced along the centerline of two adjacent plates. The contaminated gases pass through the passage between the plates, and the particles become charged and adhere to the collection plates. Collected particles are usually removed by rapping the plates and deposited in bins or hoppers at the base of the precipitator.

## Tubular Precipitators

Tubular precipitators consist of cylindrical collection electrodes with discharge electrodes located on the axis of the cylinder. The contaminated gases flow around the discharge electrode and up through the inside of the cylinders. The charged particles are collected on the grounded walls of the cylinder. The collected dust is removed from the bottom of the cylinder.

Tubular precipitators are often used for mist or fog collection or for adhesive, sticky, radioactive, or extremely toxic materials.

## Unit Collectors

Unlike central collectors, unit collectors control contamination at its source. They are small and self-contained, consisting of a fan and some form of dust collector. They are suitable for isolated, portable, or frequently moved dust-producing operations, such as bins and silos or remote belt-conveyor transfer points. Advantages of unit collectors include small space requirements, the return of collected dust to main material flow, and low initial cost. However, their dust-holding and storage capacities, servicing facilities, and maintenance periods have been sacrificed.

A number of designs are available, with capacities ranging from 200 to 2,000 $ft^3/min$ (90 to 900 L/s). There are two main types of unit collectors:

- Fabric collectors, with manual shaking or pulse-jet cleaning - normally used for fine dust
- Cyclone collectors - normally used for coarse dust

Fabric collectors are frequently used in minerals processing operations because they provide high collection efficiency and uninterrupted exhaust airflow between cleaning cycles. Cyclone collectors are used when coarser dust is generated, as in woodworking, metal grinding, or machining.

The following points should be considered when selecting a unit collector:

- Cleaning efficiency must comply with all applicable regulations.
- The unit maintains its rated capacity while accumulating large amounts of dust between cleanings.
- Simple cleaning operations do not increase the surrounding dust concentration.
- Has the ability to operate unattended for extended periods of time (for example, 8 hours).
- Automatic discharge or sufficient dust storage space to hold at least one week's accumulation.
- If renewable filters are used, they should not have to be replaced more than once a month.
- Durable.
- Quiet.

Use of unit collectors may not be appropriate if the dust-producing operations are located in an area where central exhaust systems would be practical. Dust removal and servicing requirements are expensive for many unit collectors and are more likely to be neglected than those for a single, large collector.

## Selecting a Dust Collector

Dust collectors vary widely in design, operation, effectiveness, space requirements, construction, and capital, operating, and maintenance costs. Each type has advantages and disadvantages. However, the selection of a dust collector should be based on the following general factors:

- Dust concentration and particle size - For minerals processing operations, the dust concentration can range from 0.1 to 5.0 grains (0.32 g) of dust per cubic feet of air (0.23 to 11.44 grams per standard cubic meter), and the particle size can vary from 0.5 to 100 micrometres (μm) in diameter.

- Degree of dust collection required - The degree of dust collection required depends on its potential as a health hazard or public nuisance, the plant location, the allowable emission rate, the nature of the dust, its salvage value, and so forth. The selection of a collector should be based on the efficiency required and should consider the need for high-efficiency, high-cost equipment, such as electrostatic precipitators; high-efficiency, moderate-cost equipment, such as baghouses or wet scrubbers; or lower cost, primary units, such as dry centrifugal collectors.

- Characteristics of airstream - The characteristics of the airstream can have a significant impact on collector selection. For example, cotton fabric filters cannot be used where air temperatures exceed 180 °F (82 °C). Also, condensation of steam or water vapor can blind bags. Various chemicals can attack fabric or metal and cause corrosion in wet scrubbers.

- Characteristics of dust - Moderate to heavy concentrations of many dusts (such as dust from silica sand or metal ores) can be abrasive to dry centrifugal collectors. Hygroscopic material can blind bag collectors. Sticky material can adhere to collector elements and plug passages. Some particle sizes and shapes may rule out certain types of fabric collectors. The combustible nature of many fine materials rules out the use of electrostatic precipitators.

- Methods of disposal - Methods of dust removal and disposal vary with the material, plant process, volume, and type of collector used. Collectors can unload continuously or in batches. Dry materials can create secondary dust problems during unloading and disposal that do not occur with wet collectors. Disposal of wet slurry or sludge can be an additional material-handling problem; sewer or water pollution problems can result if wastewater is not treated properly.

## Fan and Motor

The fan and motor system supplies mechanical energy to move contaminated air from the dust-producing source to a dust collector.

## Types of Fans

There are two main kinds of industrial fans:

- Centrifugal fans
- Axial-flow fans

## Centrifugal Fans

Centrifugal fans consist of a wheel or a rotor mounted on a shaft that rotates in a scroll-shaped housing. Air enters at the eye of the rotor, makes a right-angle turn, and is forced through the blades of the rotor by centrifugal force into the scroll-shaped housing. The centrifugal force imparts static pressure to the air. The diverging shape of the scroll also converts a portion of the velocity pressure into static pressure.

There are three main types of centrifugal fans:

- Radial-blade fans - Radial-blade fans are used for heavy dust loads. Their straight, radial blades do not get clogged with material, and they withstand considerable abrasion. These fans have medium tip speeds and medium noise factors.
- Backward-blade fans - Backward-blade fans operate at higher tip speeds and thus are more efficient. Since material may build up on the blades, these fans should be used after a dust collector. Although they are noisier than radial-blade fans, backward-blade fans are commonly used for large-volume dust collection systems because of their higher efficiency.
- Forward-curved-blade fans - These fans have curved blades that are tipped in the direction of rotation. They have low space requirements, low tip speeds, and a low noise factor. They are usually used against low to moderate static pressures.

## Axial-flow Fans

Axial-flow fans are used in systems that have low resistance levels. These fans move the air parallel to the fan's axis of rotation. The screw-like action of the propellers moves the air in a straight-through parallel path, causing a helical flow pattern.

The three main kinds of axial fans are-

- Propeller fans - These fans are used to move large quantities of air against very low static pressures. They are usually used for general ventilation or dilution ventilation and are good in developing up to 0.5 in. wg (124.4 Pa).
- Tube-axial fans - Tube-axial fans are similar to propeller fans except they are mounted in a tube or cylinder. Therefore, they are more efficient than propeller fans and can develop up to 3 to 4 in. wg (743.3 to 995 Pa). They are best suited for moving air containing substances such as condensible fumes or pigments.
- Vane-axial fans - Vane-axial fans are similar to tube-axial fans except air-straightening vanes are installed on the suction or discharge side of the rotor. They are easily adapted to multistaging and can develop static pressures as high as 14 to 16 in. wg (3.483 to 3.98 kPa). They are normally used for clean air only.

Fan selection

When selecting a fan, the following points should be considered:

- Volume required
- Fan static pressure
- Type of material to be handled through the fan (For example, a radial-blade fan should be

used with fibrous material or heavy dust loads, and nonsparking construction must be used with explosive or inflammable materials.)

- Type of drive arrangement, such as direct drive or belt drive
- Space requirements
- Noise levels
- Operating temperature (For example, sleeve bearings are suitable to 250 °F/121.1 °C; ball bearings to 550 °F/287.8 °C)
- Sufficient size to handle the required volume and pressure with minimum horsepower
- Need for special coatings or construction when operating in corrosive atmospheres
- Ability of fan to accommodate small changes in total pressure while maintaining the necessary air volume
- Need for an outlet damper to control airflow during cold starts (If necessary, the damper may be interlocked with the fan for a gradual start until steady-state conditions are reached).

Fan Rating Tables

After the above information is collected, the actual selection of fan size and speed is usually made from a rating table published by the fan manufacturer. This table is known as a multirating table, and it shows the complete range of capacities for a particular size of fan.

Points to note:

- The multirating table shows the range of pressures and speeds possible within the limits of the fan's construction.
- A particular fan may be available in different construction classes (identified as class I through IV) relating to its capabilities and limits.
- For a given pressure, the highest mechanical efficiency is usually found in the middle third of the volume column.
- A fan operating at a given speed can have an infinite number of ratings (pressure and volume) along the length of its characteristic curve. However, when the fan is installed in a dust collection system, the point of rating can only be at the point at which the system resistance curve intersects the fan characteristic curve.
- In a given system, a fan at a fixed speed or at a fixed blade setting can have a single rating only. This rating can be changed only be changing the fan speed, blade setting, or the system resistance.
- For a given system, an increase in exhaust volume will result in increases in static and total pressures. For example, for a 20% increase in exhaust volume in a system with 5 in. pressure loss, the new pressure loss will be $5 \times (1.20)^2 = 7.2$ in.

- For rapid estimates of probable exhaust volumes available for a given motor size, the equation for brake horsepower, as illustrated, can be useful.

Fan installation Typical fan discharge conditions Fan ratings for volume and static pressure, as described in the multirating tables, are based on the tests conducted under ideal conditions. Often, field installation creates airflow problems that reduce the fan's air delivery. The following points should be considered when installing the fan:

- Avoid installation of elbows or bends at the fan discharge, which will lower fan performance by increasing the system's resistance.
- Avoid installing fittings that may cause non-uniform flow, such as an elbow, mitred elbow, or square duct.
- Check that the fan impeller is rotating in the proper direction-clockwise or counterclockwise.
- For belt-driven fans-
    - Check that the motor sheave and fan sheave are aligned properly.
    - Check for proper belt tension.
- Check the passages between inlets, impeller blades, and inside of housing for buildup of dirt, obstructions, or trapped foreign matter.

## Electric Motors

Electric motors are used to supply the necessary energy to drive the fan.

Integral-horsepower electric motors are normally three-phase, alternating-current motors. Fractional-horsepower electric motors are normally single-phase, alternating-current motors and are used when less than 1 hp (0.75 kW) is required. Since most dust collection systems require motors with more than 1 hp (0.75 kW), only integral-horsepower motors are discussed here.

The two most common types of integral-horsepower motors used in dust collection systems are-

- Squirrel-cage motors - These motors have a constant speed and are of a nonsynchronous, induction type.
- Wound-rotor motors - These motors are also known as slip-ring motors. They are general-purpose or continuous-rated motors and are chiefly used when an adjustable-speed motor is desired.

Squirrel-cage and wound-rotor motors are further classified according to the type of enclosure they use to protect their interior windings. These enclosures fall into two broad categories:

- Open
- Totally enclosed

Drip-proof and splash-proof motors are open motors. They provide varying degrees of protection;

however, they should not be used where the air contains substances that might be harmful to the interior of the motor.

Totally enclosed motors are weather-protected with the windings enclosed. These enclosures prevent free exchange of air between the inside and the outside, but they are not airtight.

Totally enclosed, fan-cooled (TEFC) motors are another kind of totally enclosed motor. These motors are the most commonly used motors in dust collection systems. They have an integral-cooling fan outside the enclosure, but within the protective shield, that directs air over the enclosure.

Both open and totally enclosed motors are available in explosion-proof and dust-ignition-proof models to protect against explosion and fire in hazardous environments.

Motors are selected to provide sufficient power to operate fans over the full range of process conditions (temperature and flow rate).

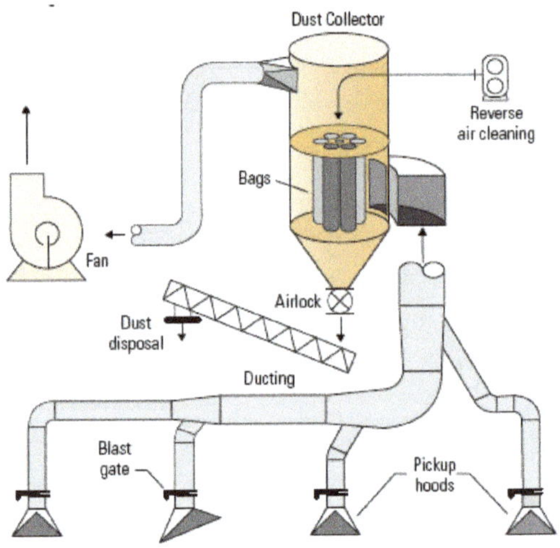

Dust Collection System Example

## Configurations

Dust collectors can be configured into one of five common types.

1. Ambient units - Ambient units are free-hanging systems for use when applications limit the use of source-capture arms or ductwork.
2. Collection booths - Collector booths require no ductwork, and allow the worker greater freedom of movement. They are often portable.
3. Downdraft tables - A downdraft table is a self-contained portable filtration system that removes harmful particulates and returns filtered air back into the facility with no external ventilation required.
4. Source collector or Portable units - Portable units are for collecting dust, mist, fumes, or smoke at the source.
5. Stationary units - An example of a stationary collector is a baghouse.

## Parameters Involved in Specifying Dust Collectors

Important parameters in specifying dust collectors include airflow the velocity of the air stream created by the vacuum producer; system power, the power of the system motor, usually specified in horsepower; storage capacity for dust and particles, and minimum particle size filtered by the unit. Other considerations when choosing a dust collection system include the temperature, moisture content, and the possibility of combustion of the dust being collected.

Systems for fine removal may only contain a single filtration system (such as a filter bag or cartridge). However, most units utilize a primary and secondary separation/filtration system. In many cases the heat or moisture content of dust can negatively affect the filter media of a baghouse or cartridge dust collector. A cyclone separator or dryer may be placed before these units to reduce heat or moisture content before reaching the filters. Furthermore, some units may have third and fourth stage filtration. All separation and filtration systems used within the unit should be specified.

A baghouse is an air pollution abatement device used to trap particulate by filtering gas streams through large fabric bags. They are typically made of glass fibers or fabric.

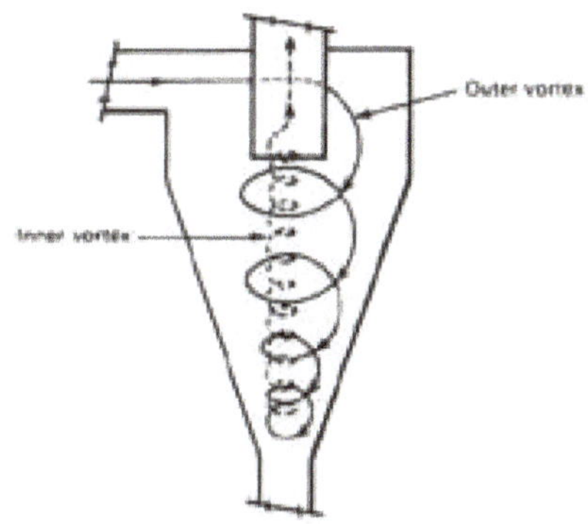


Cyclone

A cyclone separator is an apparatus for the separation, by centrifugal means, of fine particles suspended in air or gas.

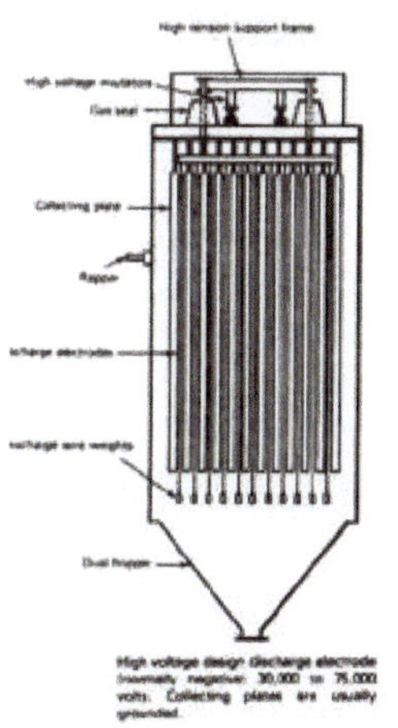

Electrostatic Precipitator

Electrostatic precipitators are a type of air cleaner, which charges particles of dust by passing dust-laden air through a strong (50-100 kV) electrostatic field. This causes the particles to be attracted to oppositely charged plates so that they can be removed from the air stream.

An impinger system is a device in which particles are removed by impacting the aerosol particles into a liquid. Modular media type units combine a variety of specific filter modules in one unit. These systems can provide solutions to many air contaminant problems. A typical system incorporates a series of disposable or cleanable pre-filters, a disposable vee-bag or cartridge filter. HEPA or carbon final filter modules can also be added. Various models are available, including free-hanging or ducted installations, vertical or horizontal mounting, and fixed or portable configurations. Filter cartridges are made out of a variety of synthetic fibers and are capable of collecting sub-micrometre particles without creating an excessive pressure drop in the system. Filter cartridges require periodic cleaning.

A wet scrubber, or venturi scrubber, is similar to a cyclone but it has an orifice unit that sprays water into the vortex in the cyclone section, collecting all of the dust in a slurry system. The water media can be recirculated and reused to continue to filter the air. Eventually the solids must be removed from the water stream and disposed of.

## Filter Cleaning Methods

Online cleaning – automatically timed filter cleaning which allows for continuous, uninterrupted dust collector operation for heavy dust operations.

Offline cleaning – filter cleaning accomplished during dust collector shut down. Practical whenever the dust loading in each dust collector cycle does not exceed the filter capacity. Allows for maximum effectiveness in dislodging and disposing of dust.

On-demand cleaning – filter cleaning initiated automatically when the filter is fully loaded, as determined by a specified drop in pressure across the media surface.

Reverse-pulse/Reverse-jet cleaning – Filter cleaning method which delivers blasts of compressed air from the clean side of the filter to dislodge the accumulated dust cake.

Impact/Rapper cleaning – Filter cleaning method in which high-velocity compressed air forced through a flexible tube results in an arbitrary rapping of the filter to dislodge the dust cake. Especially effective when the dust is extremely fine or sticky.

## Electrostatic Precipitator

An electrostatic precipitator (ESP) is a filtration device that removes fine particles, like dust and smoke, from a flowing gas using the force of an induced electrostatic charge minimally impeding the flow of gases through the unit.

In contrast to wet scrubbers which apply energy directly to the flowing fluid medium, an ESP applies energy only to the particulate matter being collected and therefore is very efficient in its consumption of energy (in the form of electricity).

Electrodes inside electrostatic precipitator

Collection electrode of electrostatic precipitator in waste incineration plant

## Invention of the Electrostatic Precipitator

The first use of corona discharge to remove particles from an aerosol was by Hohlfeld in 1824. However, it was not commercialized until almost a century later.

In 1907 Frederick Gardner Cottrell, a professor of chemistry at the University of California, Berkeley, applied for a patent on a device for charging particles and then collecting them through electrostatic attraction—the first electrostatic precipitator. Cottrell first applied the device to the collection of sulphuric acid mist and lead oxide fumes emitted from various acid-making and smelting activities. Wine-producing vineyards in northern California were being adversely affected by the lead emissions.

At the time of Cottrell's invention, the theoretical basis for operation was not understood. The operational theory was developed later in Germany, with the work of Walter Deutsch and the formation of the Lurgi company.

Cottrell used proceeds from his invention to fund scientific research through the creation of a foundation called Research Corporation in 1912, to which he assigned the patents. The intent of the organization was to bring inventions made by educators (such as Cottrell) into the commercial world for the benefit of society at large. The operation of Research Corporation is funded by royalties paid by commercial firms after commercialization occurs. Research Corporation has provided vital funding to many scientific projects: Goddard's rocketry experiments, Lawrence's cyclotron, production methods for vitamins A and $B_1$, among many others.

By a decision of the US Supreme Court, the Corporation had to be split into several entities. The Research Corporation was separated from two commercial firms making the hardware: Research-Cottrell Inc. (operating east of the Mississippi River) and Western Precipitation (operating in the western states). The Research Corporation continues to be active to this day, and the two companies formed to commercialize the invention for industrial and utility applications are still in business as well.

Electrophoresis is the term used for migration of gas-suspended charged particles in a direct-current electrostatic field. Traditional CRT television sets tend to accumulate dust on the screen because of this phenomenon (a CRT is a direct-current machine operating at about 35 kilovolts).

## Plate Precipitator

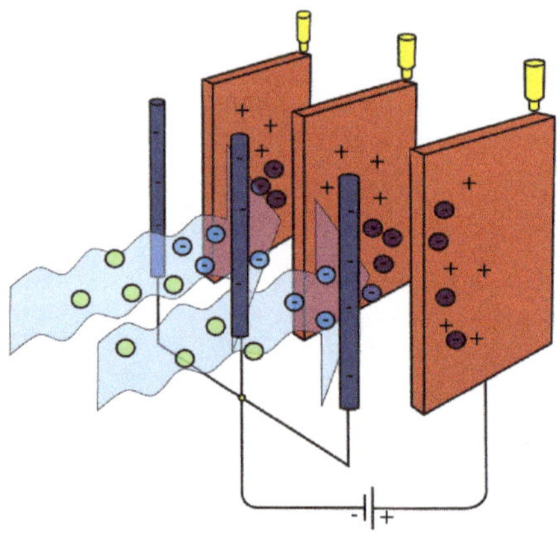

Conceptual diagram of an electrostatic precipitator

The most basic precipitator contains a row of thin vertical wires, and followed by a stack of large flat metal plates oriented vertically, with the plates typically spaced about 1 cm to 18 cm apart, depending on the application. The air stream flows horizontally through the spaces between the wires, and then passes through the stack of plates.

A negative voltage of several thousand volts is applied between wire and plate. If the applied voltage is high enough, an electric corona discharge ionizes the air around the electrodes, which then ionizes the particles in the air stream.

The ionized particles, due to the electrostatic force, are diverted towards the grounded plates. Particles build up on the collection plates and are removed from the air stream.

A two-stage design (separate charging section ahead of collecting section) has the benefit of minimizing ozone production, which would adversely affect health of personnel working in enclosed spaces. For shipboard engine rooms where gearboxes generate an oil mist, two-stage ESP's are used to clean the air, improving the operating environment and preventing buildup of flammable oil fog accumulations. Collected oil is returned to the gear lubricating system.

## Collection Efficiency (*R*)

Precipitator performance is very sensitive to two particulate properties: 1) Electrical resistivity; and 2) Particle size distribution. These properties can be measured economically and accurately in the laboratory, using standard tests. Resistivity can be determined as a function of temperature in accordance with IEEE Standard 548. This test is conducted in an air environment containing a specified moisture concentration. The test is run as a function of ascending or descending temperature, or both. Data is acquired using an average ash layer electric field of 4 kV/cm. Since relatively low applied voltage is used and no sulfuric acid vapor is present in the test environment, the values obtained indicate the maximum ash resistivity.

In an ESP, where particle charging and discharging are key functions, resistivity is an important factor that significantly affects collection efficiency. While resistivity is an important phenomenon in the inter-electrode region where most particle charging takes place, it has a particularly important effect on the dust layer at the collection electrode where discharging occurs. Particles that exhibit high resistivity are difficult to charge. But once charged, they do not readily give up their acquired charge on arrival at the collection electrode. On the other hand, particles with low resistivity easily become charged and readily release their charge to the grounded collection plate. Both extremes in resistivity impede the efficient functioning of ESPs. ESPs work best under normal resistivity conditions.

Resistivity, which is a characteristic of particles in an electric field, is a measure of a particle's resistance to transferring charge (both accepting and giving up charges). Resistivity is a function of a particle's chemical composition as well as flue gas operating conditions such as temperature and moisture. Particles can have high, moderate (normal), or low resistivity.

Bulk resistivity is defined using a more general version of Ohm's Law, as given in Equation (1) below:

$$\vec{E} = \rho \vec{j} \tag{1}$$

Where:

$E$ is the Electric field strength (V/cm);

$j$ is the Current density (A/cm$^2$); and

$\rho$ is the Resistivity (Ohm-cm)

A better way of displaying this would be to solve for resistivity as a function of applied voltage and current, as given in Equation (2) below:

$$\rho = \frac{AV}{Il} \tag{2}$$

Where:

ρ = Resistivity (Ohm-cm)

V = The applied DC potential, (Volts);

I = The measured current, (Amperes);

l = The ash layer thickness, (cm); and

A = The current measuring electrode face area, (cm$^2$).

Resistivity is the electrical resistance of a dust sample 1.0 $cm^2$ in cross-sectional area, 1.0 cm thick, and is recorded in units of ohm-cm. A method for measuring resistivity will be described in this article. The table below, gives value ranges for low, normal, and high resistivity.

| Resistivity | Range of Measurement |
|---|---|
| Low | between $10^4$ and $10^7$ ohm-cm |
| Normal | between $10^7$ and $2\times10^{10}$ ohm-cm |
| High | above $2\times10^{10}$ ohm-cm |

## Dust Layer Resistance

Resistance affects electrical conditions in the dust layer by a potential electric field (voltage drop) being formed across the layer as negatively charged particles arrive at its surface and leak their electrical charges to the collection plate. At the metal surface of the electrically grounded collection plate, the voltage is zero, whereas at the outer surface of the dust layer, where new particles and ions are arriving, the electrostatic voltage caused by the gas ions can be quite high. The strength of this electric field depends on the resistance and thickness of the dust layer.

In high-resistance dust layers, the dust is not sufficiently conductive, so electrical charges have difficulty moving through the dust layer. Consequently, electrical charges accumulate on and beneath the dust layer surface, creating a strong electric field.

Voltages can be greater than 10,000 volts. Dust particles with high resistance are held too strongly to the plate, making them difficult to remove and causing rapping problems.

In low resistance dust layers, the corona current is readily passed to the grounded collection electrode. Therefore, a relatively weak electric field, of several thousand volts, is maintained across the dust layer. Collected dust particles with low resistance do not adhere strongly enough to the collection plate. They are easily dislodged and become retained in the gas stream.

The electrical conductivity of a bulk layer of particles depends on both surface and volume factors. Volume conduction, or the motions of electrical charges through the interiors of particles, depends mainly on the composition and temperature of the particles. In the higher temperature regions, above 500 °F (260 °C), volume conduction controls the conduction mechanism. Volume conduction also involves ancillary factors, such as compression of the particle layer, particle size and shape, and surface properties.

Volume conduction is represented in the figures as a straight-line at temperatures above 500 °F (260 °C). At temperatures below about 450 °F (230 °C), electrical charges begin to flow across surface moisture and chemical films adsorbed onto the particles. Surface conduction begins to lower the resistivity values and bend the curve downward at temperatures below 500 °F (260 °C).

These films usually differ both physically and chemically from the interiors of the particles owing to adsorption phenomena. Theoretical calculations indicate that moisture films only a few molecules thick are adequate to provide the desired surface conductivity. Surface conduction on particles is closely related to surface-leakage currents occurring on electrical insulators, which have been extensively studied. An interesting practical application of surface-leakage is the determination of dew point by measurement of the current between adjacent electrodes mounted on a glass surface. A sharp rise in current signals the formation of a moisture film on the glass. This method has been used effectively for determining the marked rise in dew point, which occurs when small amounts of sulfuric acid vapor are added to an atmosphere (commercial Dewpoint Meters are available on the market).

The following discussion of normal, high, and low resistance applies to ESPs operated in a dry state; resistance is not a problem in the operation of wet ESPs because of the moisture concentration in the ESP. The relationship between moisture content and resistance is explained later in this work.

## Normal Resistivity

As stated above, ESPs work best under normal resistivity conditions. Particles with normal resistivity do not rapidly lose their charge on arrival at the collection electrode. These particles slowly leak their charge to grounded plates and are retained on the collection plates by intermolecular adhesive and cohesive forces. This allows a particulate layer to be built up and then dislodged from the plates by rapping. Within the range of normal dust resistivity (between $10^7$ and $2 \times 10^{10}$ ohm-cm), fly ash is collected more easily than dust having either low or high resistivity.

## High Resistivity

If the voltage drop across the dust layer becomes too high, several adverse effects can occur. First, the high voltage drop reduces the voltage difference between the discharge electrode and collection electrode, and thereby reduces the electrostatic field strength used to drive the gas ion-charged particles over to the collected dust layer. As the dust layer builds up, and the electrical charges accumulate on the surface of the dust layer, the voltage difference between the discharge and collection electrodes decreases. The migration velocities of small particles are especially affected by the reduced electric field strength.

Another problem that occurs with high resistivity dust layers is called back corona. This occurs when the potential drop across the dust layer is so great that corona discharges begin to appear in the gas that is trapped within the dust layer. The dust layer breaks down electrically, producing small holes or craters from which back corona discharges occur. Positive gas ions are generated within the dust layer and are accelerated toward the "negatively charged" discharge electrode. The positive ions reduce some of the negative charges on the dust layer and neutralize some of the negative ions on the "charged particles" heading toward the collection electrode. Disruptions of the normal corona process greatly reduce the ESP's collection efficiency, which in severe cases, may fall below 50% . When back corona is present, the dust particles build up on the electrodes forming a layer of insulation. Often this can not be repaired without bringing the unit offline.

The third, and generally most common problem with high resistivity dust is increased electrical sparking. When the sparking rate exceeds the "set spark rate limit," the automatic controllers limit the operating voltage of the field. This causes reduced particle charging and reduced migration velocities toward the collection electrode. High resistivity can generally be reduced by doing the following:

- Adjusting the temperature;

- Increasing moisture content;
- Adding conditioning agents to the gas stream;
- Increasing the collection surface area; and
- Using hot-side precipitators (occasionally and with foreknowledge of sodium depletion).

Thin dust layers and high-resistivity dust especially favor the formation of back corona craters. Severe back corona has been observed with dust layers as thin as 0.1 mm, but a dust layer just over one particle thick can reduce the sparking voltage by 50%. The most marked effects of back corona on the current-voltage characteristics are:

1. Reduction of the spark over voltage by as much as 50% or more;
2. Current jumps or discontinuities caused by the formation of stable back-corona craters; and
3. Large increase in maximum corona current, which just below spark over corona gap may be several times the normal current.

The figure below and to the left shows the variation in resistivity with changing gas temperature for six different industrial dusts along with three coal-fired fly ashes. The figure illustrates resistivity values measured for various chemical compounds that were prepared in the laboratory.

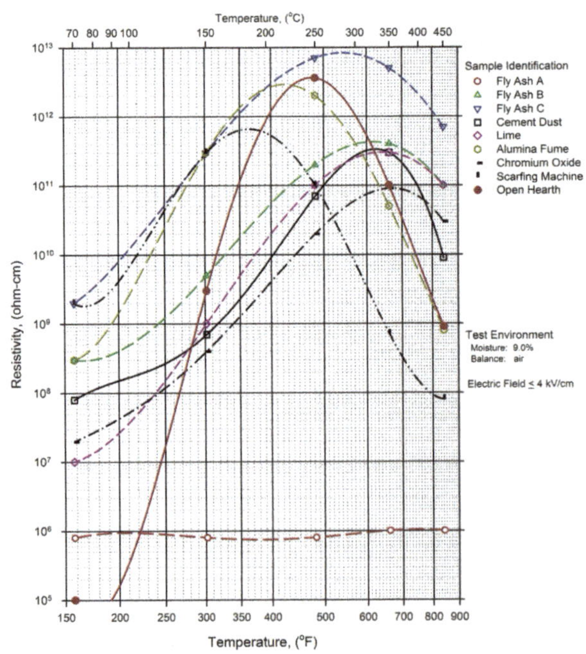


Resistivity Values of Representative Dusts and Fumes From Industrial Plants

Results for Fly Ash A were acquired in the ascending temperature mode. These data are typical for a moderate to high combustibles content ash. Data for Fly Ash B are from the same sample, acquired during the descending temperature mode.

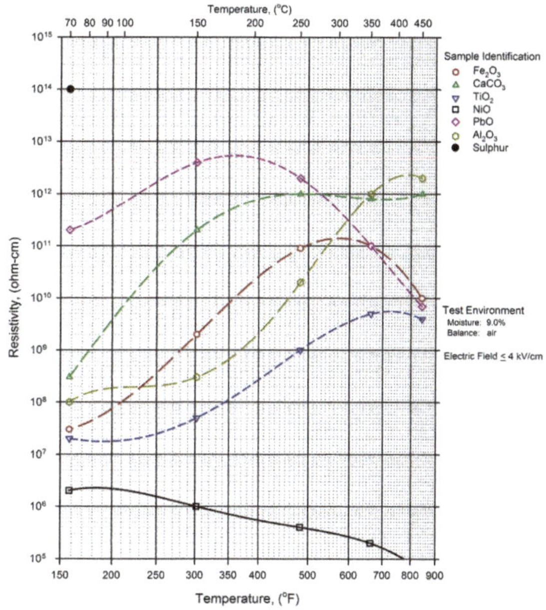

Resistivity Values of Various Chemicals and Reagents as a Function of Temperature

The differences between the ascending and descending temperature modes are due to the presence of unburned combustibles in the sample. Between the two test modes, the samples are equilibrated in dry air for 14 hours (overnight) at 850 °F (450 °C). This overnight annealing process typically removes between 60% and 90% of any unburned combustibles present in the samples. Exactly how carbon works as a charge carrier is not fully understood, but it is known to significantly reduce the resistivity of a dust.

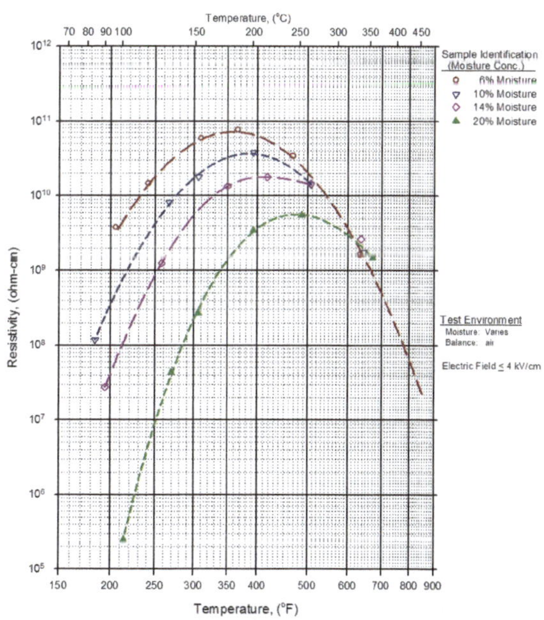

Resistivity Measured as a Function of Temperature in Varying Moisture Concentrations (Humidity)

Carbon can act, at first, like a high resistivity dust in the precipitator. Higher voltages can be required in order for corona generation to begin. These higher voltages can be problematic for the

TR-Set controls. The problem lies in onset of corona causing large amounts of current to surge through the (low resistivity) dust layer. The controls sense this surge as a spark. As precipitators are operated in spark-limiting mode, power is terminated and the corona generation cycle re-initiates. Thus, lower power (current) readings are noted with relatively high voltage readings.

The same thing is believed to occur in laboratory measurements. Parallel plate geometry is used in laboratory measurements without corona generation. A stainless steel cup holds the sample. Another stainless steel electrode weight sits on top of the sample (direct contact with the dust layer). As voltage is increased from small amounts (e.g. 20 V), no current is measured. Then, a threshold voltage level is reached. At this level, current surges through the sample... so much so that the voltage supply unit can trip off. After removal of the unburned combustibles during the above-mentioned annealing procedure, the descending temperature mode curve shows the typical inverted "V" shape one might expect.

## Low Resistivity

Particles that have low resistivity are difficult to collect because they are easily charged (very conductive) and rapidly lose their charge on arrival at the collection electrode. The particles take on the charge of the collection electrode, bounce off the plates, and become re-entrained in the gas stream. Thus, attractive and repulsive electrical forces that are normally at work at normal and higher resistivities are lacking, and the binding forces to the plate are considerably lessened. Examples of low-resistivity dusts are unburned carbon in fly ash and carbon black.

If these conductive particles are coarse, they can be removed upstream of the precipitator by using a device such as a cyclone mechanical collector.

The addition of liquid ammonia ($NH_3$) into the gas stream as a conditioning agent has found wide use in recent years. It is theorized that ammonia reacts with $H_2SO_4$ contained in the flue gas to form an ammonium sulfate compound that increases the cohesivity of the dust. This additional cohesivity makes up for the loss of electrical attraction forces.

The table below summarizes the characteristics associated with low, normal and high resistivity dusts.

The moisture content of the flue gas stream also affects particle resistivity. Increasing the moisture content of the gas stream by spraying water or injecting steam into the duct work preceding the ESP lowers the resistivity. In both temperature adjustment and moisture conditioning, one must maintain gas conditions above the dew point to prevent corrosion problems in the ESP or downstream equipment. The figure shows the effect of temperature and moisture on the resistivity of a cement dust. As the percentage of moisture in the gas stream increases from 6 to 20%, the resistivity of the dust dramatically decreases. Also, raising or lowering the temperature can decrease cement dust resistivity for all the moisture percentages represented.

The presence of $SO_3$ in the gas stream has been shown to favor the electrostatic precipitation process when problems with high resistivity occur. Most of the sulfur content in the coal burned for combustion sources converts to $SO_2$. However, approximately 1% of the sulfur converts to $SO_3$. The amount of $SO_3$ in the flue gas normally increases with increasing sulfur content of the coal. The resistivity of the particles decreases as the sulfur content of the coal increases.

| Resistivity | Range of Measurement | Precipitator Characteristics |
|---|---|---|
| Low | between $10^4$ and $10^7$ ohm-cm | 1. Normal operating voltage and current levels unless dust layer is thick enough to reduce plate clearances and cause higher current levels.<br>2. Reduced electrical force component retaining collected dust, vulnerable to high reentrainment losses.<br>3. Negligible voltage drop across dust layer.<br>4. Reduced collection performance due to (2) |
| Normal | between $10^7$ and $2 \times 10^{10}$ ohm-cm | 1. Normal operating voltage and current levels.<br>2. Negligible voltage drop across dust layer.<br>3. Sufficient electrical force component retaining collected dust.<br>4. High collection performance due to (1), (2) and (3) |
| Marginal to High | between $2 \times 10^{10}$ and $10^{12}$ ohm-cm | 1. Reduced operating voltage and current levels with high spark rates.<br>2. Significant voltage loss across dust layer.<br>3. Moderate electrical force component retaining collected dust.<br>4. Reduced collection performance due to (1) and (2) |
| High | above $10^{12}$ ohm-cm | 1. Reduced operating voltage levels; high operating current levels if power supply controller is not operating properly.<br>2. Very significant voltage loss across dust layer.<br>3. High electrical force component retaining collected dust.<br>4. Seriously reduced collection performance due to (1), (2) and probably back corona. |

Other conditioning agents, such as sulfuric acid, ammonia, sodium chloride, and soda ash (sometimes as raw trona), have also been used to reduce particle resistivity. Therefore, the chemical composition of the flue gas stream is important with regard to the resistivity of the particles to be collected in the ESP. The table below lists various conditioning agents and their mechanisms of operation.

| Conditioning Agent | Mechanism(s) of Action |
|---|---|
| Sulfur Trioxide and/or Sulfuric Acid | 1. Condensation and adsorption on fly ash surfaces.<br>2. may also increase cohesiveness of fly ash.<br>3. Reduces resistivity |
| Ammonia | Mechanism is not clear, various ones proposed;<br>1. Modifies resistivity.<br>2. Increases ash cohesiveness.<br>3. Enhances space charge effect. |
| Ammonium Sulfate | Little is known about the mechanism; claims are made for the following:<br>1. Modifies resistivity (depends upon injection temperature).<br>2. Increases ash cohesiveness.<br>3. Enhances space charge effect.<br>4. Experimental data lacking to substantiate which of these is predominant. |
| Triethylamine | Particle agglomeration claimed; no supporting data. |

| | |
|---|---|
| Sodium Compounds | 1. Natural conditioner if added with coal.<br>2. Resistivity modifier if injected into gas stream. |
| Compounds of Transition Metals | Postulated that they catalyze oxidation of $SO_2$ to $SO_3$; no definitive tests with fly ash to verify this postulation. |
| Potassium Sulfate and Sodium Chloride | In cement and lime kiln ESPs:<br>1. Resistivity modifiers in the gas stream.<br>2. NaCl - natural conditioner when mixed with coal. |

If injection of ammonium sulfate occurs at a temperature greater than about 600 °F (320 °C), dissociation into ammonia and sulfur trioxide results. Depending on the ash, $SO_2$ may preferentially interact with fly ash as $SO_3$ conditioning. The remainder recombines with ammonia to add to the space charge as well as increase cohesiveness of the ash.

More recently, it has been recognized that a major reason for loss of efficiency of the electrostatic precipitator is due to particle buildup on the charging wires in addition to the collection plates (Davidson and McKinney, 1998). This is easily remedied by making sure that the wires themselves are cleaned at the same time that the collecting plates are cleaned.

Sulfuric acid vapor ($SO_3$) enhances the effects of water vapor on surface conduction. It is physically adsorbed within the layer of moisture on the particle surfaces. The effects of relatively small amounts of acid vapor can be seen in the figure.

The inherent resistivity of the sample at 300 °F (150 °C) is $5 \times 10^{12}$ ohm-cm. An equilibrium concentration of just 1.9 ppm sulfuric acid vapor lowers that value to about $7 \times 10^9$ ohm-cm.

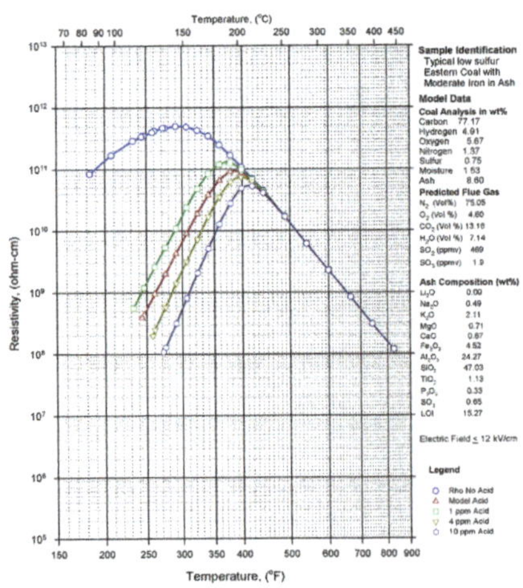


Resistivity Modeled As A Function of Environmental Conditions - Especially Sulfuric Acid Vapor

## Modern Industrial Electrostatic Precipitators

ESPs continue to be excellent devices for control of many industrial particulate emissions, including smoke from electricity-generating utilities (coal and oil fired), salt cake collection from black liquor boilers in pulp mills, and catalyst collection from fluidized bed catalytic cracker units in oil refineries to name a few. These devices treat gas volumes from several hundred thousand ACFM to 2.5 million ACFM (1,180 $m^3/s$) in the largest coal-fired boiler applications. For a coal-fired boiler the collection is usually performed downstream of the air preheater at about 160 °C (320 °F) which provides optimal resistivity of the coal-ash particles. For some difficult applications with low-sulfur fuel hot-end units have been built operating above 370 °C (698 °F).

A smokestack at coal-fired Hazelwood Power Station in Victoria, Australia emits brown smoke when its ESP is shut down

The original parallel plate–weighted wire design has evolved as more efficient (and robust) discharge electrode designs were developed, today focusing on rigid (pipe-frame) discharge electrodes to which many sharpened spikes are attached (barbed wire), maximizing corona production. Transformer-rectifier systems apply voltages of 50–100 kV at relatively high current densities. Modern controls, such as an automatic voltage control, minimize electric sparking and prevent arcing (sparks are quenched within 1/2 cycle of the TR set), avoiding damage to the components. Automatic plate-rapping systems and hopper-evacuation systems remove the collected particulate matter while on line, theoretically allowing ESPs to stay in continuous operation for years at a time.

## Electrostatic Sampling for Bioaerosols

Electrostatic precipitators can be used to sample biological airborne particles or aerosol for analysis. Sampling for bioaerosols requires precipitator designs optimised with a liquid counter electrode, which can be used to sample biological particles, e.g. viruses, directly into a small liquid volume to reduce unnecessary sample dilution.

## Wet Electrostatic Precipitator

A wet electrostatic precipitator (WESP or wet ESP) operates with water vapor saturated air streams (100% relative humidity). WESPs are commonly used to remove liquid droplets such as sulfuric acid mist from industrial process gas streams. The WESP is also commonly used where the gases are high in moisture content, contain combustible particulate, or have particles that are sticky in nature.

The preferred and most modern type of WESP is a downflow tubular design. This design allows the collected moisture and particulate to form a moving slurry that helps to keep the collection surfaces clean. Plate style and upflow design WESPs are very unreliable and should not be used in applications where particulate is sticky in nature.

## Consumer-oriented Electrostatic Air Cleaners

A portable electrostatic air cleaner marketed to consumers

Portable electrostatic air cleaner with cover removed, showing collector plates

Plate precipitators are commonly marketed to the public as air purifier devices or as a permanent replacement for furnace filters, but all have the undesirable attribute of being somewhat messy to clean. A negative side-effect of electrostatic precipitation devices is the potential production of toxic ozone and $NO_x$. However, electrostatic precipitators offer benefits over other air purifications technologies, such as HEPA filtration, which require expensive filters and can become "production sinks" for many harmful forms of bacteria.

With electrostatic precipitators, if the collection plates are allowed to accumulate large amounts of particulate matter, the particles can sometimes bond so tightly to the metal plates that vigorous washing and scrubbing may be required to completely clean the collection plates. The close spacing of the plates can make thorough cleaning difficult, and the stack of plates often cannot be easily disassembled for cleaning. One solution, suggested by several manufacturers, is to wash the collector plates in a dishwasher.

Some consumer precipitation filters are sold with special soak-off cleaners, where the entire plate array is removed from the precipitator and soaked in a large container overnight, to help loosen the tightly bonded particulates.

A study by the Canada Mortgage and Housing Corporation testing a variety of forced-air furnace filters found that ESP filters provided the best, and most cost-effective means of cleaning air using a forced-air system.

The first portable electrostatic air filter systems for homes was marketed in 1954 by Raytheon.

## Scrubber

Scrubber systems (aka Chemical Scrubbers, Gas Scrubbers) are a diverse group of air pollution control devices that can be used to remove some particulates and/or gases from industrial exhaust streams. The first air scrubber was designed to remove carbon dioxide from the air of an early submarine, the Ictineo I, a role for which they continue to be used till today. Traditionally, the term "scrubber" has referred to pollution control devices that use liquid to wash unwanted pollutants from a gas stream. Recently, the term has also been used to describe systems that inject a dry reagent or slurry into a dirty exhaust stream to "wash out" acid gases. Scrubbers are one of the primary devices that control gaseous emissions, especially acid gases. Scrubbers can also be used for heat recovery from hot gases by flue-gas condensation. They are also used for the high flows in solar, PV, or LED processes.

There are several methods to remove toxic or corrosive compounds from exhaust gas and neutralize it.

### Combustion

Combustion is sometimes the cause of harmful exhausts, but, in many cases, combustion may also be used for exhaust gas cleaning if the temperature is high enough and enough oxygen is available.

### Wet Scrubbing

The exhaust gases of combustion may contain substances considered harmful to the environment, and the scrubber may remove or neutralize those. A wet scrubber is used for cleaning air, fuel gas or other gases of various pollutants and dust particles. Wet scrubbing works via the contact of target compounds or particulate matter with the scrubbing solution. Solutions may simply be water (for dust) or solutions of reagents that specifically target certain compounds.

Process exhaust gas can also contain water-soluble toxic and/or corrosive gases like hydrochloric acid (HCl) or ammonia ($NH_3$). These can be removed very well by a wet scrubber.

Removal efficiency of pollutants is improved by increasing residence time in the scrubber or by the increase of surface area of the scrubber solution by the use of a spray nozzle, packed towers or an aspirator. Wet scrubbers may increase the proportion of water in the gas, resulting in a visible stack plume, if the gas is sent to a stack.

Wet scrubbers can also be used for heat recovery from hot gases by flue-gas condensation. In this mode, termed a condensing scrubber, water from the scrubber drain is circulated through a cooler to the nozzles at the top of the scrubber. The hot gas enters the scrubber at the bottom. If the gas temperature is above the water dew point, it is initially cooled by evaporation of water drops. Further cooling cause water vapors to condense, adding to the amount of circulating water.

The condensation of water releases significant amounts of low temperature heat (more than 2 gigajoules (560 kWh) per ton of water), that can be recovered by the cooler for e.g. district heating purposes.

Excess condensed water must continuously be removed from the circulating water.

The gas leaves the scrubber at its dew point, so even though significant amounts of water may have been removed from the cooled gas, it is likely to leave a visible stack plume of water vapor.

## Dry Scrubbing

A dry or semi-dry scrubbing system, unlike the wet scrubber, does not saturate the flue gas stream that is being treated with moisture. In some cases no moisture is added, while in others only the amount of moisture that can be evaporated in the flue gas without condensing is added. Therefore, dry scrubbers generally do not have a stack steam plume or wastewater handling/disposal requirements. Dry scrubbing systems are used to remove acid gases (such as $SO_2$ and HCl) primarily from combustion sources.

There are a number of dry type scrubbing system designs. However, all consist of two main sections or devices: a device to introduce the acid gas sorbent material into the gas stream and a particulate matter control device to remove reaction products, excess sorbent material as well as any particulate matter already in the flue gas.

Dry scrubbing systems can be categorized as dry sorbent injectors (DSIs) or as spray dryer absorbers (SDAs). Spray dryer absorbers are also called semi-dry scrubbers or spray dryers.

Dry scrubbing systems are often used for the removal of odorous and corrosive gases from wastewater treatment plant operations. The medium used is typically an activated alumina compound impregnated with materials to handle specific gases such as hydrogen sulfide. Media used can be mixed together to offer a wide range of removal for other odorous compounds such as methyl mercaptans, aldehydes, volatile organic compounds, dimethyl sulfide, and dimethyl disulfide.

Dry sorbent injection involves the addition of an alkaline material (usually hydrated lime, soda ash, or sodium bicarbonate) into the gas stream to react with the acid gases. The sorbent can be injected directly into several different locations: the combustion process, the flue gas duct (ahead of the particulate control device), or an open reaction chamber (if one exists). The acid gases react with the alkaline sorbents to form solid salts which are removed in the particulate control device. These simple systems can achieve only limited acid gas ($SO_2$ and HCl) removal efficiencies. Higher collection efficiencies can be achieved by increasing the flue gas humidity (i.e., cooling using water spray). These devices have been used on medical waste incinerators and a few municipal waste combustors.

In spray dryer absorbers, the flue gases are introduced into an absorbing tower (dryer) where the gases are contacted with a finely atomized alkaline slurry. Acid gases are absorbed by the slurry mixture and react to form solid salts which are removed by the particulate control device. The heat of the flue gas is used to evaporate all the water droplets, leaving a non-saturated flue gas to exit the absorber tower. Spray dryers are capable of achieving high (80+%) acid gas removal efficiencies. These devices have been used on industrial and utility boilers and municipal waste incinerators.

## Absorber

Many chemicals can be removed from exhaust gas also by using absorber material. The flue gas is passed through a cartridge which is filled with one or several absorber materials and has been adapted to the chemical properties of the components to be removed. This type of scrubber is sometimes also called dry scrubber. The absorber material has to be replaced after its surface is saturated.

## Mercury Removal

Mercury is a highly toxic element commonly found in coal and municipal waste. Wet scrubbers are only effective for removal of soluble mercury species, such as oxidized mercury, $Hg^{2+}$. Mercury vapor in its elemental form, $Hg^{0}$, is insoluble in the scrubber slurry and not removed. Therefore, an additional process of $Hg^{0}$ conversion is required to complete mercury capture. Usually halogens are added to the flue gas for this purpose. The type of coal burned as well as the presence of a selective catalytic reduction unit both affect the ratio of elemental to oxidized mercury in the flue gas and thus the degree to which the mercury is removed.

In July 2015, one study found that some mercury scrubbers installed on coal power plants inadvertently capture PAH (polycyclic aromatic hydrocarbons) emissions as well.

## Scrubber Waste Products

One side effect of scrubbing is that the process only moves the unwanted substance from the exhaust gases into a liquid solution, solid paste or powder form. This must be disposed of safely, if it can not be reused.

For example, mercury removal results in a waste product that either needs further processing to extract the raw mercury, or must be buried in a special hazardous wastes landfill that prevents the mercury from seeping out into the environment.

As an example of reuse, limestone-based scrubbers in coal-fired power plants can produce a synthetic gypsum of sufficient quality that can be used to manufacture drywall and other industrial products.

## Bacteria Spread

Poorly maintained scrubbers have the potential to spread disease-causing bacteria. The problem is a result of inadequate cleaning. For example, the cause of a 2005 outbreak of Legionnaires' disease in Norway was just a few infected scrubbers. The outbreak caused 10 deaths and more than 50 cases of infection.

## Baffle Spray Scrubber

Baffle spray scrubbers are a technology for air pollution control. They are very similar to spray towers in design and operation. However, in addition to using the energy provided by the spray nozzles, baffles are added to allow the gas stream to atomize some liquid as it passes over them.

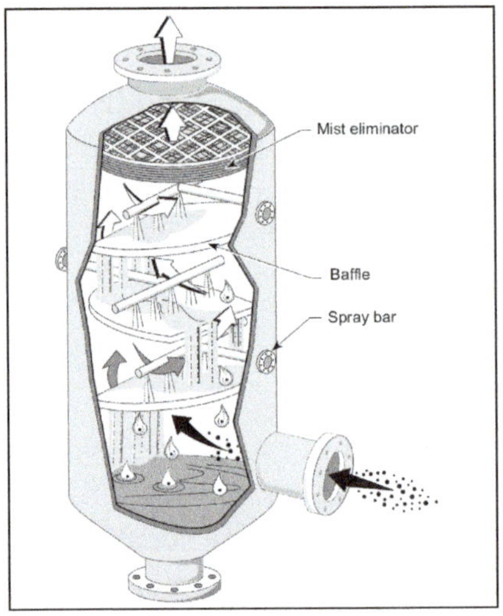


Baffle spray scrubber

A simple baffle scrubber system is shown in figure. Liquid sprays capture pollutants and also remove collected particles from the baffles. Adding baffles slightly increases the pressure drop of the system.

This type of technology is a part of the group of air pollution controls collectively referred to as wet scrubbers.

A number of wet-scrubber designs use energy from both the gas stream and liquid stream to collect pollutants. Many of these combination devices are available commercially.

A seemingly unending number of scrubber designs have been developed by changing system geometry and incorporating vanes, nozzles, and baffles.

## Particle Collection

These devices are used much the same as spray towers - to preclean or remove particles larger than 10 μm in diameter. However, they will tend to plug or corrode if particle concentration of the exhaust gas stream is high.

## Gas Collection

Even though these devices are not specifically used for gas collection, they are capable of a small amount of gas absorption because of their large wetted surface.

These devices are most commonly used as precleaners to remove large particles (>10 μm in diameter). The pressure drops across baffle scrubbers are usually low, but so are the collection efficiencies. Maintenance problems are minimal. The main problem is the buildup of solids on the baffles. Table summarizes the operating characteristics of baffle spray scrubbers.

| Operating characteristics of baffle spray scrubbers | | | | | |
|---|---|---|---|---|---|
| Pollutant | Pressure drop ($\Delta p$) | Liquid-to-gas ratio (L/G) | Liquid-inlet pressure ($p_L$) | Removal efficiency | Applications |

| | | | | | |
|---|---|---|---|---|---|
| **Gases** | 2.5-7.5 cm of water | 0.13 l/m³ (1 gal/1,000 ft³) | < 100 kPa (< 15 psig) | very low | Mining operations Incineration Chemical process industry |
| **Particles** | 1-3 in of water | | | 10 μm diameter | |

## Cyclonic Spray Scrubber

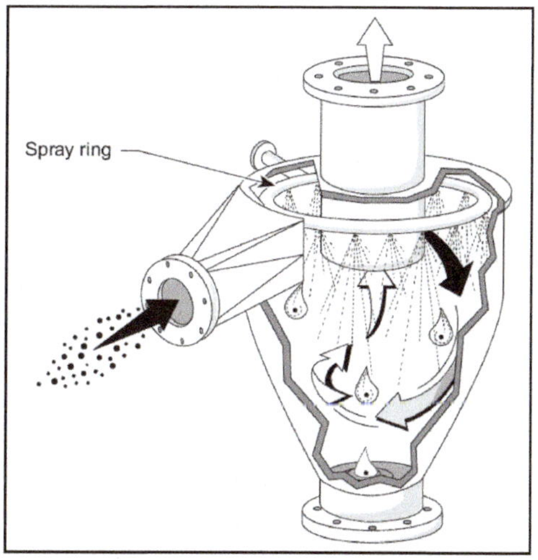


Irrigated cyclone scrubber

Cyclonic spray scrubbers are an air pollution control technology. They use the features of both the dry cyclone and the spray chamber to remove pollutants from gas streams.

Generally, the inlet gas enters the chamber tangentially, swirls through the chamber in a corkscrew motion, and exits. At the same time, liquid is sprayed inside the chamber. As the gas swirls around the chamber, pollutants are removed when they impact on liquid droplets, are thrown to the walls, and washed back down and out.

Cyclonic scrubbers are generally low- to medium-energy devices, with pressure drops of 4 to 25 cm (1.5 to 10 in) of water. Commercially available designs include the irrigated cyclone scrubber and the cyclonic spray scrubber.

In the *irrigated cyclone*, the inlet gas enters near the top of the scrubber into the water sprays. The gas is forced to swirl downward, then change directions, and return upward in a tighter spiral. The liquid droplets produced capture the pollutants, are eventually thrown to the side walls, and carried out of the collector. The “cleaned” gas leaves through the top of the chamber.

The *cyclonic spray* scrubber (figure below) forces the inlet gas up through the chamber from a bottom tangential entry. Liquid sprayed from nozzles on a center post (manifold) is directed toward the chamber walls and through the swirling gas. As in the *irrigated cyclone*, liquid captures the pollutant, is forced to the walls, and washes out. The “cleaned” gas continues upward, exiting through the straightening vanes at the top of the chamber.

This type of technology is a part of the group of air pollution controls collectively referred to as wet scrubbers.

## Particulate Collection

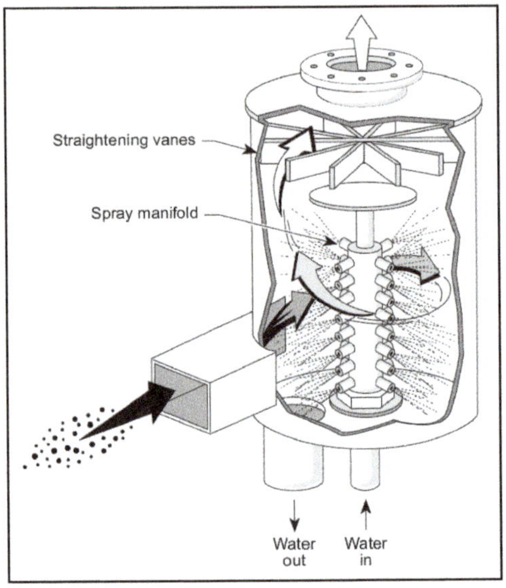

Cyclonic spray scrubber

Cyclonic spray scrubbers are more efficient than spray towers, but not as efficient as venturi scrubbers, in removing particulate from the inlet gas stream. Particulates larger than 5 μm are generally collected by impaction with 90% efficiency. In a simple spray tower, the velocity of the particulates in the gas stream is low: 0.6 to 1.5 m/s (2 to 5 ft/s).

By introducing the inlet gas tangentially into the spray chamber, the cyclonic scrubber increases gas velocities (thus, particulate velocities) to approximately 60 to 180 m/s (200 to 600 ft/s). The velocity of the liquid spray is approximately the same in both devices. This higher particulate-to-liquid relative velocity increases particulate collection efficiency for this device over that of the spray chamber. Gas velocities of 60 to 180 m/s are equivalent to those encountered in a venturi scrubber.

However, cyclonic spray scrubbers are not as efficient as venturi scrubbers because they are not capable of producing the same degree of useful turbulence.

## Gas Collection

High gas velocities through these devices reduce the gas-liquid contact time, thus reducing absorption efficiency. Cyclonic spray scrubbers are capable of effectively removing some gases; however, they are rarely chosen when gaseous pollutant removal is the only concern.

## Maintenance Problems

The main maintenance problems with cyclonic scrubbers are nozzle plugging and corrosion or erosion of the side walls of the cyclone body. Nozzles have a tendency to plug from particulates that are in the recycled liquid and/or particulates that are in the gas stream. The best solution is to install the nozzles so that they are easily accessible for cleaning or removal.

Due to high gas velocities, erosion of the side walls of the cyclone can also be a problem. Abrasion-resistant materials may be used to protect the cyclone body, especially at the inlet.

The pressure drops across cyclonic scrubbers are usually 4 to 25 cm (1.5 to 10 in) of water; therefore, they are low- to medium-energy devices and are most often used to control large-sized particulates. Relatively simple devices, they resist plugging because of their open construction. They also have the additional advantage of acting as entrainment separators because of their shape. The liquid droplets are forced to the sides of the cyclone and removed prior to exiting the vessel. Their biggest disadvantages are that they are not capable of removing submicrometer partculates and they do not efficiently absorb most pollutant gases.

Table lists typical operating characteristics of cyclonic scrubbers.

| **Operating characteristics of spray towers** | | | | | |
|---|---|---|---|---|---|
| **Pollutant** | **Pressure drop (Δp)** | **Liquid-to-gas ratio (L/G)** | **Liquid-inlet pressure ($p_L$)** | **Removal efficiency** | **Applications** |
| **Gases** | 4-25 cm of water (1.5-10 in of water) | 0.3-1.3 l/m³ (2-10 gal/1,000 ft³) | 280-2,800 kPa (40-400 psig) | Only effective for very soluble gases | Mining operations<br>Drying operations<br>Food processing<br>Foundries |
| **Particulates** | | | | 2-3 μm diameter | |

## Ejector Venturi Scrubber

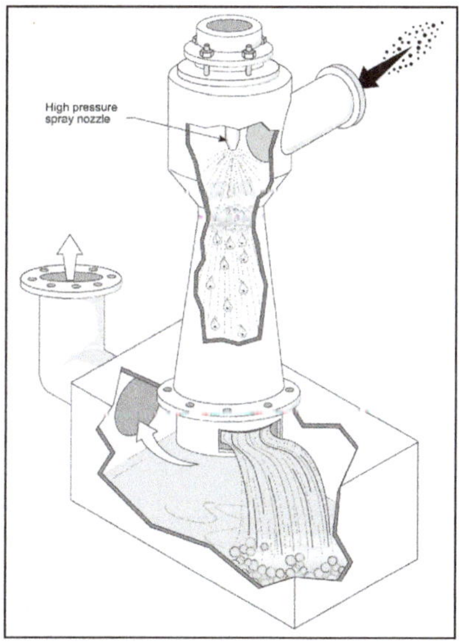


Ejector venturi scrubber

An ejector or venturi scrubber is an industrial pollution control device, usually installed on the exhaust flue gas stacks of large furnaces, but may also be used on any number of other air exhaust systems. This type of technology is a part of the group of air pollution controls collectively referred to as wet scrubbers.

Like a spray tower an ejector venturi scrubber uses a preformed spray. However, in an ejector venturi scrubber only a single nozzle is used instead of many nozzles. This nozzle operates at higher pressures and higher injection rates than those in most spray chambers. The high-pressure spray nozzle (up to 689 kPa or 100 psig) is aimed at the throat section of a venturi constriction.

The ejector venturi is unique among available scrubbing systems since it can move the process gas without the aid of a fan or blower. The liquid spray coming from the nozzle creates a partial vacuum in the side duct of the scrubber. The partial vacuum is due to the Bernoulli effect, and is similar to water aspirators used in chemistry labs. This partial vacuum can be used to move the process gas through the venturi as well as through the facility's process system. In the case of explosive or extremely corrosive atmospheres, the elimination of a fan in the system can avoid many potential problems.

The energy for the formation of scrubbing droplets comes from the injected liquid. The high pressure sprays passing through the venturi throat form numerous fine liquid droplets that provide turbulent mixing between the gas and liquid phases. Very high liquid-injection rates are used to provide the gas-moving capability and higher collection efficiencies. As with other types of venturis, a means of separating entrained liquid from the gas stream must be installed. Entrainment separators are commonly used to remove remaining small droplets.

## Particle Collection

Ejector venturis are effective in removing particles larger than 1.0 μm in diameter. These scrubbers are not used on submicrometer-sized particles unless the particles are condensable [Gilbert, 1977]. Particle collection occurs primarily by impaction as the exhaust gas (from the process) passes through the spray.

The turbulence that occurs in the throat area also causes the particles to contact the wet droplets and be collected. Particle collection efficiency increases with an increase in nozzle pressure and/or an increase in the liquid-to-gas ratio. Increases in either of these two operating parameters will also result in an increase in pressure drop for a given system. Therefore, an increase in pressure drop also increases particle collection efficiency. Ejector venturis operate at higher L/G ratios than most other particulate scrubbers (i.e., 7 to 13 $l/m^3$ compared to 0.4-2.7 $l/m^3$ for most other designs).

## Gas Collection

Ejector venturis have a short gas-liquid contact time because the exhaust gas velocities through the vessel are very high. This short contact time limits the absorption efficiency of the system. Although ejector venturis are not used primarily for gas removal, they can be effective if the gas is very soluble or if a very reactive scrubbing reagent is used. In these instances, removal efficiencies of as high as 95% can be achieved [Gilbert, 1977].

## Maintenance Problems

Ejector venturis are subject to abrasion problems in the high-velocity areas - nozzle and throat. Both must be constructed of wear-resistant materials because of the high liquid injection rates and nozzle pressures. Maintaining the pump that recirculates liquid is also very important. In addition, the high gas velocities necessitate the use of entrainment separators to prevent excessive liquid carryover. The separators should be easily accessible or removable so that they can be cleaned if plugging occurs.

Because of their open design and the fact that they do not require a fan, ejector venturis are capable of handling a wide range of corrosive and/or sticky particles. However, they are not very effective in removing submicrometer particles. They have an advantage in being able to handle small, medium and large exhaust flows. They can be used singly or in multiple stages of two or more in series, depending on the specific application.

Multiple-stage systems have been used where extremely high collection efficiency of particles or gaseous pollutants was necessary. Multiple-stage systems provide increased gas-liquid contact time, thus increasing absorption efficiency.

Table lists the operating parameters for ejector venturis.

**Operating characteristics of ejector venturis**

| Pollutant | Pressure drop (Δp) | Liquid-to-gas ratio (L/G) | Liquid-inlet pressure ($p_L$) | Removal efficiency | Applications |
|---|---|---|---|---|---|
| Gases | 1.3–13 cm of water | 7-13 l/m³ | 100-830 kPa | 95% for very soluble gases | Pulp and paper industry<br>Chemical process industry<br>Food industry<br>Metals processing industry |
| Particles | 0.5-5 in of water | 50-100 gal/1,000 ft³ | 15-120 psig | 1 μm diameter | |

## Sewage Treatment

Wastewater treatment plant in Massachusetts, United States

Sewage treatment is the process of removing contaminants from wastewater, primarily from household sewage. It includes physical, chemical, and biological processes to remove these contaminants and produce environmentally safer treated wastewater (or treated effluent). A by-product of sewage treatment is usually a semi-solid waste or slurry, called sewage sludge, that has to undergo further treatment before being suitable for disposal or land application.

Sewage treatment may also be referred to as wastewater treatment, although the latter is a broader term which can also be applied to purely industrial wastewater. For most cities, the sewer system will also carry a proportion of industrial effluent to the sewage treatment plant which has usually received pretreatment at the factories themselves to reduce the pollutant load. If the sewer system is a combined sewer then it will also carry urban runoff (stormwater) to the sewage treatment plant. Sewage water can travel towards treatment plants via piping and in a flow aided by gravity and pumps. The first part of filtration of sewage typically includes a bar screen to filter solids and large objects which are then collected in dumpsters and disposed of in landfills. Fat and grease will also be removed before the primary treatment of sewage.

## Terminology

The term "sewage treatment plant" (or "sewage treatment works" in some countries) is nowadays often replaced with the term "wastewater treatment plant".

Sewage can be treated close to where the sewage is created, which may be called a "decentralized" system or even an "on-site" system (in septic tanks, biofilters or aerobic treatment systems). Alternatively, sewage can be collected and transported by a network of pipes and pump stations to a municipal treatment plant. This is called a "centralized" system.

## Origins of Sewage

Sewage is generated by residential, institutional, commercial and industrial establishments. It includes household waste liquid from toilets, baths, showers, kitchens, and sinks draining into sewers. In many areas, sewage also includes liquid waste from industry and commerce. The separation and draining of household waste into greywater and blackwater is becoming more common in the developed world, with treated greywater being permitted to be used for watering plants or recycled for flushing toilets.

## Sewage Mixing with Rainwater

Sewage may include stormwater runoff or urban runoff. Sewerage systems capable of handling storm water are known as combined sewer systems. This design was common when urban sewerage systems were first developed, in the late 19th and early 20th centuries. Combined sewers require much larger and more expensive treatment facilities than sanitary sewers. Heavy volumes of storm runoff may overwhelm the sewage treatment system, causing a spill or overflow. Sanitary sewers are typically much smaller than combined sewers, and they are not designed to transport stormwater. Backups of raw sewage can occur if excessive infiltration/inflow (dilution by stormwater and/or groundwater) is allowed into a sanitary sewer system. Communities that have urbanized in the mid-20th century or later generally have built separate systems for sewage (sanitary sewers) and stormwater, because precipitation causes widely varying flows, reducing sewage treatment plant efficiency.

As rainfall travels over roofs and the ground, it may pick up various contaminants including soil particles and other sediment, heavy metals, organic compounds, animal waste, and oil and grease. Some jurisdictions require stormwater to receive some level of treatment before being discharged directly into waterways. Examples of treatment processes used for stormwater include retention basins, wetlands, buried vaults with various kinds of media filters, and vortex separators (to remove coarse solids).

## Industrial Effluent

In highly regulated developed countries, industrial effluent usually receives at least pretreatment if not full treatment at the factories themselves to reduce the pollutant load, before discharge to the sewer. This process is called industrial wastewater treatment. The same does not apply to many developing countries where industrial effluent is more likely to enter the sewer if it exists, or even the receiving water body, without pretreatment.

Industrial wastewater may contain pollutants which cannot be removed by conventional sewage treatment. Also, variable flow of industrial waste associated with production cycles may upset the population dynamics of biological treatment units, such as the activated sludge process.

## Process Steps

Sewage collection and treatment is typically subject to local, state and federal regulations and standards.

Treating wastewater has the aim to produce an effluent that will do as little harm as possible when discharged to the surrounding environment, thereby preventing pollution compared to releasing untreated wastewater into the environment.

Sewage treatment generally involves three stages, called primary, secondary and tertiary treatment.

- *Primary treatment* consists of temporarily holding the sewage in a quiescent basin where heavy solids can settle to the bottom while oil, grease and lighter solids float to the surface. The settled and floating materials are removed and the remaining liquid may be discharged or subjected to secondary treatment. Some sewage treatment plants that are connected to a combined sewer system have a bypass arrangement after the primary treatment unit. This means that during very heavy rainfall events, the secondary and tertiary treatment systems can be bypassed to protect them from hydraulic overloading, and the mixture of sewage and stormwater only receives primary treatment.

- *Secondary treatment* removes dissolved and suspended biological matter. Secondary treatment is typically performed by indigenous, water-borne micro-organisms in a managed habitat. Secondary treatment may require a separation process to remove the micro-organisms from the treated water prior to discharge or tertiary treatment.

- *Tertiary treatment* is sometimes defined as anything more than primary and secondary treatment in order to allow rejection into a highly sensitive or fragile ecosystem (estuaries, low-flow rivers, coral reefs,...). Treated water is sometimes disinfected chemically or physically (for example, by lagoons and microfiltration) prior to discharge into a stream, river, bay, lagoon or wetland, or it can be used for the irrigation of a golf course, green way or park. If it is sufficiently clean, it can also be used for groundwater recharge or agricultural purposes.

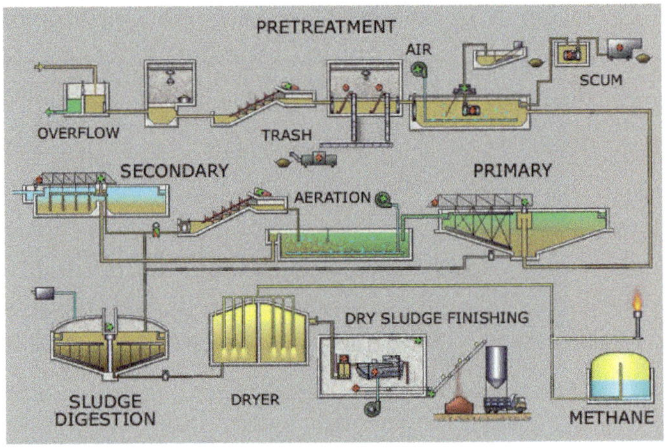


Simplified process flow diagram for a typical large-scale treatment plant

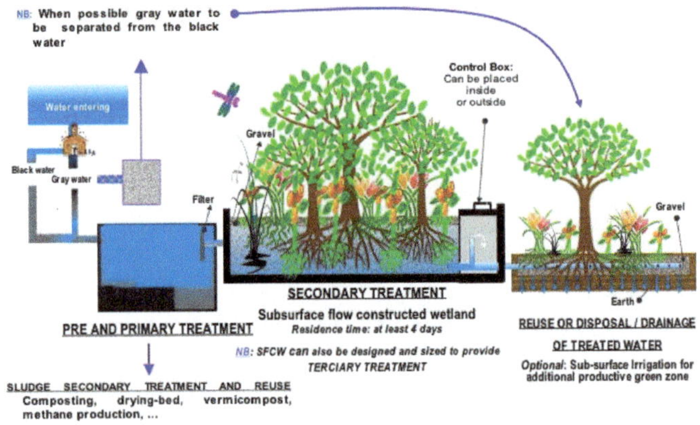


Process flow diagram for a typical treatment plant via subsurface flow constructed wetlands (SFCW)

## Pretreatment

Pretreatment removes all materials that can be easily collected from the raw sewage before they damage or clog the pumps and sewage lines of primary treatment clarifiers. Objects commonly removed during pretreatment include trash, tree limbs, leaves, branches, and other large objects.

The influent in sewage water passes through a bar screen to remove all large objects like cans, rags, sticks, plastic packets etc. carried in the sewage stream. This is most commonly done with an automated mechanically raked bar screen in modern plants serving large populations, while in smaller or less modern plants, a manually cleaned screen may be used. The raking action of a mechanical bar screen is typically paced according to the accumulation on the bar screens and/or flow rate. The solids are collected and later disposed in a landfill, or incinerated. Bar screens or mesh screens of varying sizes may be used to optimize solids removal. If gross solids are not removed, they become entrained in pipes and moving parts of the treatment plant, and can cause substantial damage and inefficiency in the process.

## Grit Removal

Pretreatment may include a sand or grit channel or chamber, where the velocity of the incoming sewage is adjusted to allow the settlement of sand, grit, stones, and broken glass. These particles are removed because they may damage pumps and other equipment. For small sanitary sewer systems, the grit chambers may not be necessary, but grit removal is desirable at larger plants. Grit chambers come in 3 types: horizontal grit chambers, aerated grit chambers and vortex grit chambers. The process is called sedimentation.

## Flow Equalization

Clarifiers and mechanized secondary treatment are more efficient under uniform flow conditions. Equalization basins may be used for temporary storage of diurnal or wet-weather flow peaks. Basins provide a place to temporarily hold incoming sewage during plant maintenance and a means of diluting and distributing batch discharges of toxic or high-strength waste which might otherwise inhibit biological secondary treatment (including portable toilet waste, vehicle holding tanks, and septic tank pumpers). Flow equalization basins require variable discharge control, typically include provisions for bypass and cleaning, and may also include aerators. Cleaning may be easier if the basin is downstream of screening and grit removal.

## Fat and Grease Removal

In some larger plants, fat and grease are removed by passing the sewage through a small tank where skimmers collect the fat floating on the surface. Air blowers in the base of the tank may also be used to help recover the fat as a froth. Many plants, however, use primary clarifiers with mechanical surface skimmers for fat and grease removal.

## Primary Treatment

Primary treatment tanks in Oregon, USA.

In the primary sedimentation stage, sewage flows through large tanks, commonly called "pre-settling basins", "primary sedimentation tanks" or "primary clarifiers". The tanks are used to settle sludge while grease and oils rise to the surface and are skimmed off. Primary settling tanks are usually equipped with mechanically driven scrapers that continually drive the collected sludge towards a hopper in the base of the tank where it is pumped to sludge treatment facilities. Grease and oil from the floating material can sometimes be recovered for saponification (soap making).

## Secondary Treatment

Secondary treatment is designed to substantially degrade the biological content of the sewage which are derived from human waste, food waste, soaps and detergent. The majority of municipal plants treat the settled sewage liquor using aerobic biological processes. To be effective, the biota require both oxygen and food to live. The bacteria and protozoa consume biodegradable soluble organic contaminants (e.g. sugars, fats, organic short-chain carbon molecules, etc.) and bind much of the less soluble fractions into floc. Secondary treatment systems are classified as *fixed-film* or *suspended-growth* systems.

- Fixed-film or attached growth systems include trickling filters, bio-towers, and rotating biological contactors, where the biomass grows on media and the sewage passes over its surface. The fixed-film principle has further developed into Moving Bed Biofilm Reactors (MBBR) and Integrated Fixed-Film Activated Sludge (IFAS) processes. An MBBR system typically requires a smaller footprint than suspended-growth systems.
- Suspended-growth systems include activated sludge, where the biomass is mixed with the sewage and can be operated in a smaller space than trickling filters that treat the same amount of water. However, fixed-film systems are more able to cope with drastic changes in the amount of biological material and can provide higher removal rates for organic material and suspended solids than suspended growth systems.

## Secondary Sedimentation

Secondary clarifier at a rural treatment plant.

Some secondary treatment methods include a secondary clarifier to settle out and separate biological floc or filter material grown in the secondary treatment bioreactor.

## List of Process Types

- Activated sludge
- Aerated lagoon
- Aerobic granulation
- Constructed wetland
- Membrane bioreactor

- Rotating biological contactor
- Sequencing batch reactor
- Trickling filter

To use less space, treat difficult waste, and intermittent flows, a number of designs of hybrid treatment plants have been produced. Such plants often combine at least two stages of the three main treatment stages into one combined stage. In the UK, where a large number of wastewater treatment plants serve small populations, package plants are a viable alternative to building a large structure for each process stage. In the US, package plants are typically used in rural areas, highway rest stops and trailer parks.

## Tertiary Treatment

The purpose of tertiary treatment is to provide a final treatment stage to further improve the effluent quality before it is discharged to the receiving environment (sea, river, lake, wet lands, ground, etc.). More than one tertiary treatment process may be used at any treatment plant. If disinfection is practised, it is always the final process. It is also called "effluent polishing."

## Filtration

Sand filtration removes much of the residual suspended matter. Filtration over activated carbon, also called *carbon adsorption,* removes residual toxins.

## Lagoons or Ponds

A sewage treatment plant and lagoon in Everett, Washington, United States.

Lagoons or ponds provide settlement and further biological improvement through storage in large man-made ponds or lagoons. These lagoons are highly aerobic and colonization by native macrophytes, especially reeds, is often encouraged. Small filter-feeding invertebrates such as *Daphnia* and species of *Rotifera* greatly assist in treatment by removing fine particulates.

## Biological Nutrient Removal

Biological nutrient removal (BNR) is regarded by some as a type of secondary treatment process, and by others as a tertiary (or "advanced") treatment process.

Wastewater may contain high levels of the nutrients nitrogen and phosphorus. Excessive release to the environment can lead to a buildup of nutrients, called eutrophication, which can in turn encourage the overgrowth of weeds, algae, and cyanobacteria (blue-green algae). This may cause an algal bloom, a rapid growth in the population of algae. The algae numbers are unsustainable and eventually most of them die. The decomposition of the algae by bacteria uses up so much of the oxygen in the water that most or all of the animals die, which creates more organic matter for the bacteria to decompose. In addition to causing deoxygenation, some algal species produce toxins that contaminate drinking water supplies. Different treatment processes are required to remove nitrogen and phosphorus.

## Nitrogen Removal

Nitrogen is removed through the biological oxidation of nitrogen from ammonia to nitrate (nitrification), followed by denitrification, the reduction of nitrate to nitrogen gas. Nitrogen gas is released to the atmosphere and thus removed from the water.

Nitrification itself is a two-step aerobic process, each step facilitated by a different type of bacteria. The oxidation of ammonia ($NH_3$) to nitrite ($NO_2^-$) is most often facilitated by *Nitrosomonas* spp. ("nitroso" referring to the formation of a nitroso functional group). Nitrite oxidation to nitrate ($NO_3^-$), though traditionally believed to be facilitated by *Nitrobacter* spp. (nitro referring the formation of a nitro functional group), is now known to be facilitated in the environment almost exclusively by *Nitrospira* spp.

Denitrification requires anoxic conditions to encourage the appropriate biological communities to form. It is facilitated by a wide diversity of bacteria. Sand filters, lagooning and reed beds can all be used to reduce nitrogen, but the activated sludge process (if designed well) can do the job the most easily. Since denitrification is the reduction of nitrate to dinitrogen (molecular nitrogen) gas, an electron donor is needed. This can be, depending on the waste water, organic matter (from feces), sulfide, or an added donor like methanol. The sludge in the anoxic tanks (denitrification tanks) must be mixed well (mixture of recirculated mixed liquor, return activated sludge (RAS), and raw influent) e.g. by using submersible mixers in order to achieve the desired denitrification.

Sometimes the conversion of toxic ammonia to nitrate alone is referred to as tertiary treatment.

Over time, different treatment configurations have evolved as denitrification has become more sophisticated. An initial scheme, the Ludzack-Ettinger Process, placed an anoxic treatment zone before the aeration tank and clarifier, using the return activated sludge (RAS) from the clarifier as a nitrate source. Influent wastewater (either raw or as effluent from primary clarification) serves as the electron source for the facultative bacteria to metabolize carbon, using the inorganic nitrate as a source of oxygen instead of dissolved molecular oxygen. This denitrification scheme was naturally limited to the amount of soluble nitrate present in the RAS. Nitrate reduction was limited because RAS rate is limited by the performance of the clarifier.

The "Modified Ludzak-Ettinger Process" (MLE) is an improvement on the original concept, for it recycles mixed liquor from the discharge end of the aeration tank to the head of the anoxic tank to provide a consistent source of soluble nitrate for the facultative bacteria. In this instance, raw wastewater continues to provide the electron source, and sub-surface mixing maintains the bacteria in contact with both electron source and soluble nitrate in the absence of dissolved oxygen.

Many sewage treatment plants use centrifugal pumps to transfer the nitrified mixed liquor from the aeration zone to the anoxic zone for denitrification. These pumps are often referred to as *Internal Mixed Liquor Recycle* (IMLR) pumps. IMLR may be 200% to 400% the flow rate of influent wastewater (Q.) This is in addition to Return Activated Sludge (RAS) from secondary clarifiers, which may be 100% of Q. (Therefore, the hydraulic capacity of the tanks in such a system should handle at least 400% of annual average design flow (AADF.) At times, the raw or primary effluent wastewater must be carbon-supplemented by the addition of methanol, acetate, or simple food waste (molasses, whey, plant starch) to improve the treatment efficiency. These carbon additions should be accounted for in the design of a treatment facility's organic loading.

Further modifications to the MLE were to come: Bardenpho and Biodenipho processes include additional anoxic and oxidative processes to further polish the conversion of nitrate ion to molecular nitrogen gas. Use of an anaerobic tank following the initial anoxic process allows for luxury uptake of phosphorus by bacteria, thereby biologically reducing orthophosphate ion in the treated wastewater. Even newer improvements, such as Anammox Process, interrupt the formation of nitrate at the nitrite stage of nitrification, shunting nitrite-rich mixed liquor activated sludge to treatment where nitrite is then converted to molecular nitrogen gas, saving energy, alkalinity, and secondary carbon sourcing. Anammox™ (ANaerobic AMMonia OXidation) works by artificially extending detention time and preserving denitrifiying bacteria through the use of substrate added to the mixed liquor and continuously recycled from it prior to secondary clarification. Many other proprietary schemes are being deployed, including DEMON™, Sharon-ANAMMOX™, ANITA-Mox™, and DeAmmon™. The bacteria Brocadia anammoxidans can remove ammonium from waste water through anaerobic oxidation of ammonium to hydrazine, a form of rocket fuel.

## Phosphorus Removal

Every adult human excretes between 200 and 1000 grams of phosphorus annually. Studies of United States sewage in the late 1960s estimated mean per capita contributions of 500 grams in urine and feces, 1000 grams in synthetic detergents, and lesser variable amounts used as corrosion and scale control chemicals in water supplies. Source control via alternative detergent formulations has subsequently reduced the largest contribution, but the content of urine and feces will remain unchanged. Phosphorus removal is important as it is a limiting nutrient for algae growth in many fresh water systems. . It is also particularly important for water reuse systems where high phosphorus concentrations may lead to fouling of downstream equipment such as reverse osmosis.

Phosphorus can be removed biologically in a process called enhanced biological phosphorus removal. In this process, specific bacteria, called polyphosphate-accumulating organisms (PAOs), are selectively enriched and accumulate large quantities of phosphorus within their cells (up to 20 percent of their mass). When the biomass enriched in these bacteria is separated from the treated water, these biosolids have a high fertilizer value.

Phosphorus removal can also be achieved by chemical precipitation, usually with salts of iron (e.g. ferric chloride), aluminum (e.g. alum), or lime. This may lead to excessive sludge production as hydroxides precipitates and the added chemicals can be expensive. Chemical phosphorus removal requires significantly smaller equipment footprint than biological removal, is easier to operate and is often more reliable than biological phosphorus removal. Another method for phosphorus removal is to use granular laterite.

Once removed, phosphorus, in the form of a phosphate-rich sewage sludge, may be dumped in a landfill or used as fertilizer. In the latter case, the treated sewage sludge is also sometimes referred to as biosolids.

## Disinfection

The purpose of disinfection in the treatment of waste water is to substantially reduce the number of microorganisms in the water to be discharged back into the environment for the later use of drinking, bathing, irrigation, etc. The effectiveness of disinfection depends on the quality of the water being treated (e.g., cloudiness, pH, etc.), the type of disinfection being used, the disinfectant dosage (concentration and time), and other environmental variables. Cloudy water will be treated less successfully, since solid matter can shield organisms, especially from ultraviolet light or if contact times are low. Generally, short contact times, low doses and high flows all militate against effective disinfection. Common methods of disinfection include ozone, chlorine, ultraviolet light, or sodium hypochlorite. Chloramine, which is used for drinking water, is not used in the treatment of waste water because of its persistence. After multiple steps of disinfection, the treated water is ready to be released back into the water cycle by means of the nearest body of water or agriculture. Afterwards, the water can be transferred to reserves for everyday human uses.

Chlorination remains the most common form of waste water disinfection in North America due to its low cost and long-term history of effectiveness. One disadvantage is that chlorination of residual organic material can generate chlorinated-organic compounds that may be carcinogenic or harmful to the environment. Residual chlorine or chloramines may also be capable of chlorinating organic material in the natural aquatic environment. Further, because residual chlorine is toxic to aquatic species, the treated effluent must also be chemically dechlorinated, adding to the complexity and cost of treatment.

Ultraviolet (UV) light can be used instead of chlorine, iodine, or other chemicals. Because no chemicals are used, the treated water has no adverse effect on organisms that later consume it, as may be the case with other methods. UV radiation causes damage to the genetic structure of bacteria, viruses, and other pathogens, making them incapable of reproduction. The key disadvantages of UV disinfection are the need for frequent lamp maintenance and replacement and the need for a highly treated effluent to ensure that the target microorganisms are not shielded from the UV radiation (i.e., any solids present in the treated effluent may protect microorganisms from the UV light). In the United Kingdom, UV light is becoming the most common means of disinfection because of the concerns about the impacts of chlorine in chlorinating residual organics in the wastewater and in chlorinating organics in the receiving water. Some sewage treatment systems in Canada and the US also use UV light for their effluent water disinfection.

Ozone ($O_3$) is generated by passing oxygen ($O_2$) through a high voltage potential resulting in a third oxygen atom becoming attached and forming $O_3$. Ozone is very unstable and reactive and oxidizes most organic material it comes in contact with, thereby destroying many pathogenic microorganisms. Ozone is considered to be safer than chlorine because, unlike chlorine which has to be stored on site (highly poisonous in the event of an accidental release), ozone is generated on-site as needed. Ozonation also produces fewer disinfection by-products than chlorination. A disadvantage of ozone disinfection is the high cost of the ozone generation equipment and the requirements for special operators.

## Fourth Treatment Stage

Micropollutants such as pharmaceuticals, ingredients of household chemicals, chemicals used in small businesses or industries, environmental persistent pharmaceutical pollutant (EPPP) or pesticides may not be eliminated in the conventional treatment process (primary, secondary and tertiary treatment) and therefore lead to water pollution. Although concentrations of those substances and their decompostion products are quite low, there is still a chance to harm aquatic organisms. For pharmaceuticals, the following substances have been identified as "toxicologically relevant": substances with endocrine disrupting effects, genotoxic substances and substances that enhance the development of bacterial resistances. They mainly belong to the group of environmental persistent pharmaceutical pollutants. Techniques for elimination of micropollutants via a fourth treatment stage during sewage treatment are being tested in Germany, Switzerland and the Netherlands. However, since those techniques are still costly, they are not yet applied on a regular basis. Such process steps mainly consist of activated carbon filters that adsorb the micropollutants. Ozone can also be applied as an oxidative method. Also the use of enzymes such as the enzyme laccase is under investigation. A new concept which could provide an energy-efficient treatment of micropollutants could be the use of laccase secreting fungi cultivated at a wastewater treatment plant to degrade micropollutants and at the same time to provide enzymes at a cathode of a microbial biofuel cells. Microbial biofuel cells are investigated for their property to treat organic matter in wastewater.

To reduce pharmaceuticals in water bodies, also "source control" measures are under investigation, such as innovations in drug development or more responsible handling of drugs.

## Odor Control

Odors emitted by sewage treatment are typically an indication of an anaerobic or "septic" condition. Early stages of processing will tend to produce foul-smelling gases, with hydrogen sulfide being most common in generating complaints. Large process plants in urban areas will often treat the odors with carbon reactors, a contact media with bio-slimes, small doses of chlorine, or circulating fluids to biologically capture and metabolize the noxious gases. Other methods of odor control exist, including addition of iron salts, hydrogen peroxide, calcium nitrate, etc. to manage hydrogen sulfide levels.

High-density solids pumps are suitable for reducing odors by conveying sludge through hermetic closed pipework.

## Energy Requirements

For conventional sewage treatment plants, around 30 percent of the annual operating costs is usually required for energy. The energy requirements vary with type of treatment process as well as wastewater load. For example, constructed wetlands have a lower energy requirement than activated sludge plants, as less energy is required for the aeration step. Sewage treatment plants that produce biogas in their sewage sludge treatment process with anaerobic digestion can produce enough energy to meet most of the energy needs of the sewage treatment plant itself.

In conventional secondary treatment processes, most of the electricity is used for aeration, pumping systems and equipment for the dewatering and drying of sewage sludge. Advanced wastewater treatment plants, e.g. for nutrient removal, require more energy than plants that only achieve primary or secondary treatment.

## Sludge Treatment and Disposal

The sludges accumulated in a wastewater treatment process must be treated and disposed of in a safe and effective manner. The purpose of digestion is to reduce the amount of organic matter and the number of disease-causing microorganisms present in the solids. The most common treatment options include anaerobic digestion, aerobic digestion, and composting. Incineration is also used, albeit to a much lesser degree.

Sludge treatment depends on the amount of solids generated and other site-specific conditions. Composting is most often applied to small-scale plants with aerobic digestion for mid-sized operations, and anaerobic digestion for the larger-scale operations.

The sludge is sometimes passed through a so-called pre-thickener which de-waters the sludge. Types of pre-thickeners include centrifugal sludge thickeners rotary drum sludge thickeners and belt filter presses. Dewatered sludge may be incinerated or transported offsite for disposal in a landfill or use as an agricultural soil amendment.

## Environment Aspects

The outlet of the Karlsruhe sewage treatment plant flows into the Alb.

Many processes in a wastewater treatment plant are designed to mimic the natural treatment processes that occur in the environment, whether that environment is a natural water body or the ground. If not overloaded, bacteria in the environment will consume organic contaminants, although this will reduce the levels of oxygen in the water and may significantly change the overall ecology of the receiving water. Native bacterial populations feed on the organic contaminants, and the numbers of disease-causing microorganisms are reduced by natural environmental conditions such as predation or exposure to ultraviolet radiation. Consequently, in cases where the receiving environment provides a high level of dilution, a high degree of wastewater treatment may not be required. However, recent evidence has demonstrated that very low levels of specific contaminants in wastewater, including hormones (from animal husbandry and residue from human hormonal contraception methods) and synthetic materials such as phthalates that mimic hormones in their action, can have an unpredictable adverse impact on the natural biota and potentially on humans if the water is re-used for drinking water. In the US and EU, uncontrolled discharges of wastewater to the environment are not permitted under law, and strict water quality requirements are to be met, as clean drinking water is essential. A significant threat in the coming decades will be the increasing uncontrolled discharges of wastewater within rapidly developing countries.

## Effects on Biology

Sewage treatment plants can have multiple effects on nutrient levels in the water that the treated sewage flows into. These nutrients can have large effects on the biological life in the water in contact with the effluent. Stabilization ponds (or sewage treatment ponds) can include any of the following:

- Oxidation ponds, which are aerobic bodies of water usually 1–2 meters in depth that receive effluent from sedimentation tanks or other forms of primary treatment.
- Dominated by algae
- Polishing ponds are similar to oxidation ponds but receive effluent from an oxidation pond or from a plant with an extended mechanical treatment.
- Dominated by zooplankton
- Facultative lagoons, raw sewage lagoons, or sewage lagoons are ponds where sewage is added with no primary treatment other than coarse screening. These ponds provide effective treatment when the surface remains aerobic; although anaerobic conditions may develop near the layer of settled sludge on the bottom of the pond.
- Anaerobic lagoons are heavily loaded ponds.
- Dominated by bacteria
- Sludge lagoons are aerobic ponds, usually 2 to 5 meters in depth, that receive anaerobically digested primary sludge, or activated secondary sludge under water.
- Upper layers are dominated by algae

Phosphorus limitation is a possible result from sewage treatment and results in flagellate-dominated plankton, particularly in summer and fall.

A phytoplankton study found high nutrient concentrations linked to sewage effluents. High nutrient concentration leads to high chlorophyll a concentrations, which is a proxy for primary production in marine environments. High primary production means high phytoplankton populations and most likely high zooplankton populations, because zooplankton feed on phytoplankton. However, effluent released into marine systems also leads to greater population instability.

The planktonic trends of high populations close to input of treated sewage is contrasted by the bacterial trend. In a study of *Aeromonas* spp. in increasing distance from a wastewater source, greater change in seasonal cycles was found the furthest from the effluent. This trend is so strong that the furthest location studied actually had an inversion of the *Aeromonas* spp. cycle in comparison to that of fecal coliforms. Since there is a main pattern in the cycles that occurred simultaneously at all stations it indicates seasonal factors (temperature, solar radiation, phytoplankton) control of the bacterial population. The effluent dominant species changes from *Aeromonas caviae* in winter to *Aeromonas sobria* in the spring and fall while the inflow dominant species is *Aeromonas caviae*, which is constant throughout the seasons.

## Treated Sewage Reuse

With suitable technology, it is possible to reuse sewage effluent for drinking water, although this is usually only done in places with limited water supplies, such as Windhoek and Singapore.

In Israel, about 50 percent of agricultural water use (total use was 1 billion cubic metres in 2008) is provided through reclaimed sewer water. Future plans call for increased use of treated sewer water as well as more desalination plants.

## Sewage Treatment in Developing Countries

Few reliable figures exist on the share of the wastewater collected in sewers that is being treated in the world. A global estimate by UNDP and UN-Habitat is that 90% of all wastewater generated is released into the environment untreated. In many developing countries the bulk of domestic and industrial wastewater is discharged without any treatment or after primary treatment only.

In Latin America about 15 percent of collected wastewater passes through treatment plants (with varying levels of actual treatment). In Venezuela, a below average country in South America with respect to wastewater treatment, 97 percent of the country's sewage is discharged raw into the environment. In Iran, a relatively developed Middle Eastern country, the majority of Tehran's population has totally untreated sewage injected to the city's groundwater. However, the construction of major parts of the sewage system, collection and treatment, in Tehran is almost complete, and under development, due to be fully completed by the end of 2012. In Isfahan, Iran's third largest city, sewage treatment was started more than 100 years ago.

Only few cities in sub-Saharan Africa have sewer-based sanitation systems, let alone wastewater treatment plants, an exception being South Africa and – until the late 1990s- Zimbabwe. Instead, most urban residents in sub-Saharan Africa rely on on-site sanitation systems without sewers, such as septic tanks and pit latrines, and faecal sludge management in these cities is an enormous challenge.

## History

Basic sewer systems were used for waste removal in ancient Mesopotamia, where vertical shafts carried the waste away into cesspools. Similar systems existed in the Indus Valley civilization in modern-day India and in Ancient Crete and Greece. In the Middle Ages the sewer systems built by the Romans fell into disuse and waste was collected into cesspools that were periodically emptied by workers known as 'rakers' who would often sell it as fertilizer to farmers outside the city.

The Great Stink of 1858 stimulated research into the problem of sewage treatment.
In this caricature in *The Times*, Michael Faraday reports to *Father Thames* on the state of the river.

Modern sewage systems were first built in the mid-nineteenth century as a reaction to the exacerbation of sanitary conditions brought on by heavy industrialization and urbanization. Due to the contaminated water supply, cholera outbreaks occurred in 1832, 1849 and 1855 in London, killing tens of thousands of people. This, combined with the Great Stink of 1858, when the smell of untreated human waste in the River Thames became overpowering, and the report into sanitation reform of the Royal Commissioner Edwin Chadwick, led to the Metropolitan Commission of Sewers appointing Sir Joseph Bazalgette to construct a vast underground sewage system for the safe removal of waste. Contrary to Chadwick's recommendations, Bazalgette's system, and others later built in Continental Europe, did not pump the sewage onto farm land for use as fertilizer; it was simply piped to a natural waterway away from population centres, and pumped back into the environment.

## Early Attempts

One of the first attempts at diverting sewage for use as a fertilizer in the farm was made by the cotton mill owner James Smith in the 1840s. He experimented with a piped distribution system initially proposed by James Vetch that collected sewage from his factory and pumped it into the outlying farms, and his success was enthusiastically followed by Edwin Chadwick and supported by organic chemist Justus von Liebig.

The idea was officially adopted by the Health of Towns Commission, and various schemes (known as sewage farms) were trialled by different municipalities over the next 50 years. At first, the heavier solids were channeled into ditches on the side of the farm and were covered over when full, but soon flat-bottomed tanks were employed as reservoirs for the sewage; the earliest patent was taken out by William Higgs in 1846 for "tanks or reservoirs in which the contents of sewers and drains from cities, towns and villages are to be collected and the solid animal or vegetable matters therein contained, solidified and dried..." Improvements to the design of the tanks included the introduction of the horizontal-flow tank in the 1850s and the radial-flow tank in 1905. These tanks had to be manually de-sludged periodically, until the introduction of automatic mechanical de-sludgers in the early 1900s.

The precursor to the modern septic tank was the cesspool in which the water was sealed off to prevent contamination and the solid waste was slowly liquified due to anaerobic action; it was invented by L.H Mouras in France in the 1860s. Donald Cameron, as City Surveyor for Exeter patented an improved version in 1895, which he called a 'septic tank'; septic having the meaning of 'bacterial'. These are still in worldwide use, especially in rural areas unconnected to large-scale sewage systems.

## Chemical Treatment

Sir Edward Frankland, a distinguished chemist, who demonstrated the possibility of chemically treating sewage in the 1870s.

It was not until the late 19th century that it became possible to treat the sewage by chemically breaking it down through the use of microorganisms and removing the pollutants. Land treatment was also steadily becoming less feasible, as cities grew and the volume of sewage produced could no longer be absorbed by the farmland on the outskirts.

Sir Edward Frankland conducted experiments at the Sewage Farm in Croydon, England, during the 1870s and was able to demonstrate that filtration of sewage through porous gravel produced a nitrified effluent (the ammonia was converted into nitrate) and that the filter remained unclogged over long periods of time. This established the then revolutionary possibility of biological treatment of sewage using a contact bed to oxidize the waste. This concept was taken up by the chief chemist for the London Metropolitan Board of Works, William Libdin, in 1887:

> ...in all probability the true way of purifying sewage...will be first to separate the sludge, and then turn into neutral effluent... retain it for a sufficient period, during which time it should be fully aerated, and finally discharge it into the stream in a purified condition. This is indeed what is aimed at and imperfectly accomplished on a sewage farm.

From 1885 to 1891 filters working on this principle were constructed throughout the UK and the idea was also taken up in the US at the Lawrence Experiment Station in Massachusetts, where Frankland's work was confirmed. In 1890 the LES developed a 'trickling filter' that gave a much more reliable performance.

Contact beds were developed in Salford, Lancashire and by scientists working for the London City Council in the early 1890s. According to Christopher Hamlin, this was part of a conceptual revolution that replaced the philosophy that saw "sewage purification as the prevention of decomposition with one that tried to facilitate the biological process that destroy sewage naturally."

Contact beds were tanks containing the inert substance, such as stones or slate, that maximized the surface area available for the microbial growth to break down the sewage. The sewage was held in the tank until it was fully decomposed and it was then filtered out into the ground. This method quickly became widespread, especially in the UK, where it was used in Leicester, Sheffield, Manchester and Leeds. The bacterial bed was simultaneously developed by Joseph Corbett as Borough Engineer in Salford and experiments in 1905 showed that his method was superior in that greater volumes of sewage could be purified better for longer periods of time than could be achieved by the contact bed.

The Royal Commission on Sewage Disposal published its eighth report in 1912 that set what became the international standard for sewage discharge into rivers; the '20:30 standard', which allowed 20 mg Biochemical oxygen demand and 30 mg suspended solid per litre.

## Aerated Lagoon

An aerated lagoon or aerated basin is a holding and/or treatment pond provided with artificial aeration to promote the biological oxidation of wastewaters. There are many other biological processes for treatment of wastewaters, for example activated sludge, trickling filters, rotating biological contactors and biofilters. They all have in common the use of oxygen (or air) and microbial action to biotreat the pollutants in wastewaters.

## Types of Aerated Lagoons or Basins

- Suspension mixed lagoons, where there is less energy provided by the aeration equipment to keep the sludge in suspension.
- Facultative lagoons, where there is insufficient energy provided by the aeration equipment to keep the sludge in suspension and solids settle to the lagoon floor. The biodegradable solids in the settled sludge then degrade as in an anaerobic lagoon.

## Suspension Mixed Lagoons

Suspension mixed lagoons flow through activated sludge systems where the effluent has the same composition as the mixed liquor in the lagoon. Typically the sludge will have a residence time or sludge age of 1 to 5 days. This means that the chemical oxygen demand (COD) removed is relatively little and the effluent is therefore unacceptable for discharge into receiving waters. The objective of the lagoon is therefore to act as a biologically assisted flocculator which converts the soluble biodegradable organics in the influent to a biomass which is able to settle as a sludge. Usually the effluent is then put in a second pond where the sludge can settle. The effluent can then be removed from the top with a low COD, while the sludge accumulates on the floor and undergoes anaerobic stabilisation.

## Methods of Aerating Lagoons or Basins

There are many methods for aerating a lagoon or basin:

- Motor-driven submerged or floating jet aerators
- Motor-driven floating surface aerators
- Motor-driven fixed-in-place surface aerators
- Injection of compressed air through submerged diffusers

## Floating Surface Aerators

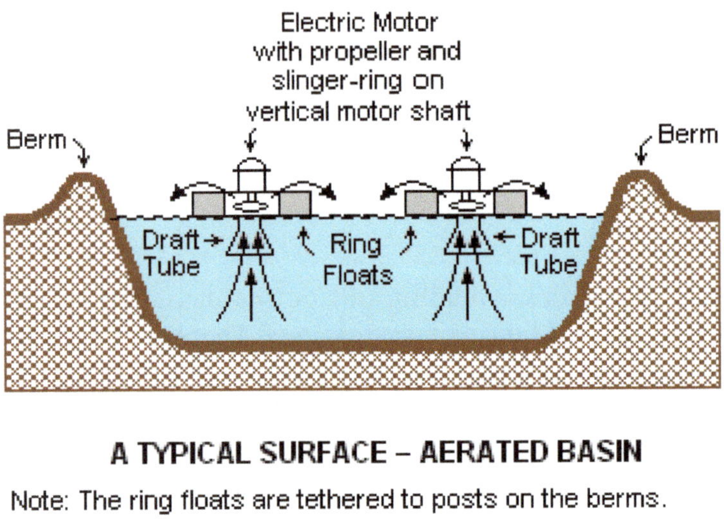


A Typical Surface-Aerated Basin (using motor-driven floating aerators)

Ponds or basins using floating surface aerators achieve 80 to 90% removal of BOD with retention times of 1 to 10 days. The ponds or basins may range in depth from 1.5 to 5.0 metres.

In a surface-aerated system, the aerators provide two functions: they transfer air into the basins required by the biological oxidation reactions, and they provide the mixing required for dispersing the air and for contacting the reactants (that is, oxygen, wastewater and microbes). Typically, the floating high speed surface aerators are rated to deliver the amount of air equivalent to 1 to 1.2 kg $O_2$/kWh. However, they do not provide as good mixing as is normally achieved in activated sludge systems and therefore aerated basins do not achieve the same performance level as activated sludge units.

With low speed surface aerators SOTE (Standard Oxygen Transfer Efficiency) is higher thanks to better mixing capacity. This mixing capacity of an impeller depends highly on the impeller diameter. Low speed surface aerator present such high diameter. Therefore SOTE for low speed surface aerators is about 2 to 2.5 kg $O_2$/kWh. This is why low speed surface aerators are mostly used in sewage or industrial treatment as WWTP are bigger and sparing energy (and money) becomes very interesting.

Biological oxidation processes are sensitive to temperature and, between 0 °C and 40 °C, the rate of biological reactions increase with temperature. Most surface aerated vessels operate at between 4 °C and 32 °C.

## Submerged Diffused Aeration

Submerged diffused air is essentially a form of a diffuser grid inside a lagoon. There are two main types of submerged diffused aeration systems for lagoon applications: floating lateral and submerged lateral. Both these systems utilize fine or medium bubble diffusers to provide aeration and mixing to the process water. The diffusers can be suspended slightly above the lagoon floor or may rest on the bottom. Flexible airline or weighted air hose supplies air to the diffuser unit from the air lateral (either floating or submerged).

## Activated Sludge

Activated sludge tank at Beckton sewage treatment plant, UK - the white bubbles are due to the diffused air aeration system

The activated sludge process is a process for treating sewage and industrial wastewaters using air and a biological floc composed of bacteria and protozoa.

PM : Its important to realize that the name 'Activated Sludge' is reserved for aerobic processes, where in anaerobic processes also active bacteria are in function and mostly in very high volumetric rates.

## Purpose

In a sewage (or industrial wastewater) treatment plant, the activated sludge process is a biological process that can be used for one or several of the following purposes: oxidizing carbonaceous biological matter, oxidizing nitrogenous matter: mainly ammonium and nitrogen in biological matter, removing nutrients (nitrogen and phosphorus).

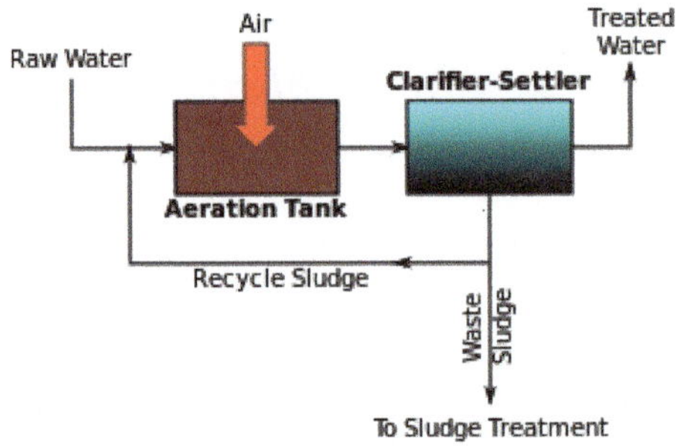


A generalized, schematic diagram of an activated sludge process.

Activated sludge addition (seeding) to a pilot scale membrane bioreactor in Germany

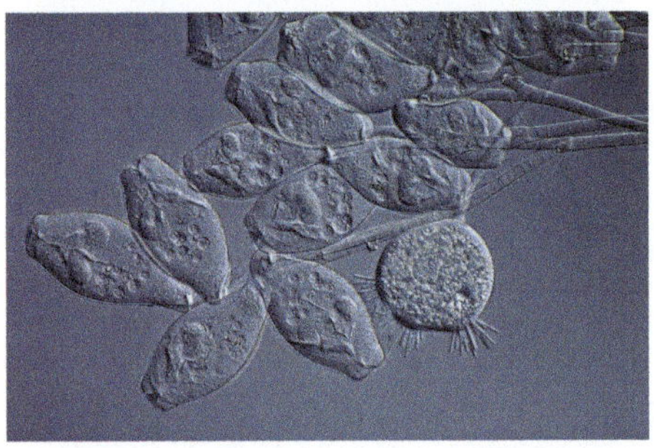

Activated sludge under the microscope

## Process Description

The general arrangement of an activated sludge process for removing carbonaceous pollution includes the following items:

- Aeration tank where air (or oxygen) is injected in the mixed liquor.
- Settling tank (usually referred to as "final clarifier" or "secondary settling tank") to allow the biological flocs (the sludge blanket) to settle, thus separating the biological sludge from the clear treated water.

Treatment of nitrogenous matter or phosphorus involves additional steps where the mixed liquor is left in anoxic condition (meaning that there is no residual dissolved oxygen) and anaerobic zones (meaning that next to dissolved oxygen also Nitrite and Nitrate are absent).

## Bioreactor and Final Clarifier

The process involves air or oxygen being introduced into a mixture of screened, and primary treated sewage or industrial wastewater (wastewater) combined with organisms to develop a biological floc which reduces the organic content of the sewage. This material, which in healthy sludge is a brown floc, is largely composed of saprotrophic bacteria but also has an important protozoan flora component mainly composed of amoebae, Spirotrichs, Peritrichs including Vorticellids and a range of other filter-feeding species. Other important constituents include motile and sedentary Rotifers. In poorly managed activated sludge, a range of mucilaginous filamentous bacteria can develop including *Sphaerotilus natans* which produces a sludge that is difficult to settle and can result in the sludge blanket decanting over the weirs in the settlement tank to severely contaminate the final effluent quality. This material is often described as sewage fungus but true fungal communities are relatively uncommon.

The combination of wastewater and biological mass is commonly known as *mixed liquor*. In all activated sludge plants, once the wastewater has received sufficient treatment, the mixed liquor is settled in settling tanks and the treated or purified supernatant is run off to discharge or sometimes undergoes further treatment before discharge or reuse. The majority of the settled material, the sludge, is returned to the head of the aeration system to re-seed the new wastewater entering the tank. This fraction of the floc is called *return activated sludge* (RAS).

The space required for a sewage treatment plant can be reduced by using a membrane bioreactor to remove some wastewater from the mixed liquor prior to treatment. This results in a more concentrated waste product that can then be treated using the activated sludge process.

Many sewage treatment plants use axial flow pumps to transfer nitrified mixed liquor from the aeration zone to the anoxic zone for denitrification. These pumps are often referred to as internal mixed liquor recycle pumps (IMLR pumps). The raw sewage, the RAS, and the nitrified mixed liquor are mixed by submersible mixers in the anoxic zones in order to achieve denitrification.

## Sludge Production

Activated sludge is also the name given to the active biological material produced by activated sludge plants. Excess sludge is called "surplus activated sludge" or "waste activated sludge" or "excess aerobic biomass" and is removed from the treatment process to keep the ratio of biomass to food supplied in the wastewater in balance. This sewage sludge can be mixed with primary sludge from the primary clarifiers and undergoes further sludge treatment for example by anaerobic digestion, followed by thickening, dewatering, composting and land application. Today more often primary and secondary sludge are thickened separately before mixed towards digestion.

The amount of sewage sludge produced from the activated sludge process is directly proportional to the amount of wastewater treated, but more specific proportional to the load of pollutants in kg COD or BOD. The total sludge production consists of the sum of primary sludge from the primary sedimentation tanks as well as waste activated sludge from the bioreactors. A typical activated sludge process for domestic wastewater produces about 70–100 $g/m^3$ of waste activated sludge (that is g of dry solids produced per $m^3$ of wastewater treated). A value of 80 $g/m^3$ is regarded as being typical. In addition, about 110–170 g/m³ of primary sludge is produced in the primary sedimentation tanks which most - but not all - of the activated sludge process configurations use.

A variant of the activated sludge process is the Nereda process where aerobic granular sludge is developed by applying specific process conditions that favour slow growing organisms.

## Process Control

The general method to do this is to monitor sludge blanket level, the biomass concentration (MLSS in g/l), the Sludge Volume after 30 minutes settling (SV30 in ml/l), SVI (Sludge Volume Index), MCRT (Mean Cell Residence Time) or SRT (Sludge Retention Time in days), F/M (Food to Microorganism), as well as the biota of the activated sludge and the major nutrients DO (Dissolved oxygen), nitrogen both in reduced (NH4-N) and oxidezed forms (NO2-N and NO3-N), phosphorus, COD (Chemical oxygen demand) and BOD (Biochemical oxygen demand).

In the reactor/aerator + clarifier system:

- The sludge blanket is measured from the bottom of the clarifier to the level of settled solids in the clarifier's water column; this, in large plants, can be done up to three times a day.
- The SVI is the volume of settled sludge in milliliters occupied by 1 gram of dry sludge solids after 30 minutes of settling in a 1000 milliliter graduated cylinder.
- The MCRT is the total mass (lbs) of mixed liquor suspended solids in the aerator and clarifier divided by the mass flow rate (lbs/day) of mixed liquor suspended solids leaving as WAS and final effluent.
- The F/M is the ratio of food fed to the microorganisms each day to the mass of microorganisms held under aeration. Specifically, it is the amount of BOD fed to the aerator (lbs/day) divided by the amount (lbs) of Mixed Liquor Volatile Suspended Solids (MLVSS) under aeration. Note: Some references use MLSS (Mixed Liquor Suspended Solids) for expedience, but MLVSS is considered more accurate for the measure of microorganisms. Again, due to expedience, COD is generally used, in lieu of BOD, as COD is a rather easy and reliable analyzing method and as BOD takes five days for results.

Based on these control methods, the amount of settled solids in the mixed liquor can be varied by wasting activated sludge (WAS) or returning activated sludge (RAS).

## Types of Plants

There are a variety of types of activated sludge plants. These include:

## Package Plants

There are a wide range of types of package plants, often serving small communities or industrial plants that may use hybrid treatment processes often involving the use of aerobic sludge to treat the incoming (and preference settled) sewage. In such plants the primary settlement stage of treatment may be omitted (PM I do not agree (JR) removal of solids is important for package or attached biofilm processes). In these plants, a biotic floc is created which provides the required substrate. Package plants are designed and fabricated by specialty engineering firms in dimensions that allow for their transportation to the job site in public highways, typically width and height of 12 by 12 feet. Length varies with capacity with larger plants being fabricated in pieces and welded on site. Steel is preferred over synthetic materials (e.g., plastic) for its durability.

Package plants are commonly variants of extended aeration, to promote the 'fit & forget' approach required for small communities without dedicated operational staff. There are various standards to assist with their design.

## Oxidation Ditch

In some areas, where more land is available, sewage is treated in large round or oval ditches with one or more horizontal aerators typically called brush or disc aerators which drive the mixed liquor around the ditch and provide aeration. These are oxidation ditches, often referred to by manufacturer's trade names such as Pasveer, Orbal or Carrousel. They have the advantage that they are relatively easy to maintain and are resilient to shock loads that often occur in smaller communities (i.e. at breakfast time and in the evening).

Oxidation ditches are installed commonly as 'fit & forget' technology, with typical design parameters of a hydraulic retention time of 24 – 48 hours, and a Sludge Retention Time or sludge age of 12 – 20 days. This compares with nitrifying activated sludge plants having a retention time of 8 hours, and a sludge age of 8 – 12 days.

## Deep Shaft/Vertical Treatment

Where land is in short supply sewage may be treated by injection of oxygen into a pressured return sludge stream which is injected into the base of a deep columnar tank buried in the ground. Such shafts may be up to 100 metres deep and are filled with sewage liquor. As the sewage rises the oxygen forced into solution by the pressure at the base of the shaft breaks out as molecular oxygen providing a highly efficient source of oxygen for the activated sludge biota. The rising oxygen and injected return sludge provide the physical mechanism for mixing of the sewage and sludge. Mixed sludge and sewage is decanted at the surface and separated into supernatant and sludge components. The efficiency of deep shaft treatment can be high.

Surface aerators are commonly quoted as having an aeration efficiency of 0.5 - 1.5 kg $O_2$/kWh, diffused aeration as 1.5 - 2.5 kg $O_2$/kWh. Deep Shaft claims 5 – 8 kg $O_2$/kWh.

However, the costs of construction are high. Deep Shaft has seen the greatest uptake in Japan, because of the land area issues. Deep Shaft was developed by ICI, as a spin-off from their Pruteen process. In the UK it is found at three sites: Tilbury, Anglian water, treating a wastewater with a high industrial contribution; Southport, United Utilities, because of land space issues; and Billingham, ICI, again treating industrial effluent, and built (after the Tilbury shafts) by ICI to help the agent sell more.

DeepShaft is a patented, licensed, process. The licensee has changed several times and currently (2015) Noram Engineering sells it.

## Surface-aerated Basins

Most biological oxidation processes for treating industrial wastewaters have in common the use of oxygen (or air) and microbial action. Surface-aerated basins achieve 80 to 90% removal of BOD with retention times of 1 to 10 days. The basins may range in depth from 1.5 to 5.0 metres and utilize motor-driven aerators floating on the surface of the wastewater.

In an aerated basin system, the aerators provide two functions: they transfer air into the basins required by the biological oxidation reactions, and they provide the mixing required for dispersing the air and for contacting the reactants (that is, oxygen, wastewater and microbes). Typically, the floating surface aerators are rated to deliver the amount of air equivalent to 1.8 to 2.7 kg $O_2$/kWh. However, they do not provide as good mixing as is normally achieved in activated sludge systems and therefore aerated basins do not achieve the same performance level as activated sludge units.

Biological oxidation processes are sensitive to temperature and, between 0 °C and 40 °C, the rate of biological reactions increase with temperature. Most surface aerated vessels operate at between 4 °C and 32 °C.

## Sequencing Batch Reactors (SBRs)

Sequencing batch reactors (SBRs) treat wastewater in batches within the same vessel. This means that the bioreactor (and the specific aerobic, anoxic and anaerobic circumstances) and final clarifier are not separated in space but in a timed sequence. The installation consists of at least two identically equipped tanks with a common inlet, which can be switched between them. While one tank is in settle/decant mode the other is aerating and filling.

## Aeration Methods

### Diffused Aeration

Sewage liquor is run into deep tanks with diffuser grid aeration systems that are attached to the floor. These are like the diffused airstone used in tropical fish tanks but on a much larger scale. Air is pumped through the blocks and the curtain of bubbles formed both oxygenates the liquor and also provides the necessary mixing action. Where capacity is limited or the sewage is unusually strong or difficult to treat, oxygen may be used instead of air. Typically, the air is generated by some type of blower or compressor.

### Surface Aerators (Cones)

Vertically mounted tubes of up to 1-metre diameter extending from just above the base of a deep concrete tank to just below the surface of the sewage liquor. A typical shaft might be 10 metres high. At the surface end, the tube is formed into a cone with helical vanes attached to the inner surface. When the tube is rotated, the vanes spin liquor up and out of the cones drawing new sewage liquor from the base of the tank. In many works, each cone is located in a separate cell that can be isolated from the remaining cells if required for maintenance. Some works may have two cones to a cell and some large works may have 4 cones per cell.

## Pure Oxygen Aeration

Pure oxygen activated sludge aeration systems are sealed-tank reactor vessels with surface aerator type impellers mounted within the tanks at the oxygen atmosphere-mixed liquor surface interface. The amount of oxygen entrainment, or DO (Dissolved Oxygen), can be controlled by a weir adjusted level control, and a vent gas oxygen controlled oxygen feed valve. Oxygen is generated on site by cryogenic distillation of air, pressure swing adsorption, or other methods. These systems are used where wastewater plant space is at a premium and high sewage throughput is required as high energy costs are involved in purifying oxygen.

## Recent Developments

A new development of the activated sludge process is the Nereda process which produces a granular sludge that settles very well (the sludge volume index is reduced from 200-300 to 40 mL/g). A new process reactor system is created to take advantage of this quick settling sludge and is integrated into the aeration tank instead of having a separate unit outside. About 30 Nereda wastewater treatment plants worldwide are operational, under construction or under design, varying in size from 5,000 up to 858,000 person equivalent.

## History

The Davyhulme Sewage Works Laboratory, where the activated sludge process was developed in the early 20th century.

The activated sludge process was discovered in 1913 in the United Kingdom by two engineers, Edward Ardern and W.T. Lockett, who were conducting research for the Manchester Corporation Rivers Department at Davyhulme Sewage Works. This development led to arguably the single most significant improvement in public health and the environment during the course of the century.

In 1912, Dr. Gilbert Fowler, a scientist at the University of Manchester, observed experiments being conducted at the Lawrence Experiment Station at Massachusetts involving the aeration of sewage in a bottle that had been coated with algae. Fowler's engineering colleagues, Ardern and Lockett, experimented on treating sewage in a draw-and-fill reactor, which produced a highly treated effluent. They aerated the waste-water continuously for about a month and were able to achieve a complete nitrification of the sample material. Believing that the sludge had been activated (in a similar manner to activated carbon) the process was named *activated sludge*. Not until much later was it realized that what had actually occurred was a means to concentrate biological organisms, decoupling the liquid retention time (ideally, low, for a compact treatment system) from the solids retention time (ideally, fairly high, for an effluent low in $BOD_5$ and ammonia.)

Their results were published in their seminal 1914 paper, and the first full-scale continuous-flow system was installed at Worcester two years later. In the aftermath of the First World War the new treatment method spread rapidly, especially to the USA, Denmark, Germany and Canada. By the late 1930s, the activated sludge treatment became a well-known biological wastewater treatment process in those countries where sewer systems and sewage treatment plants were common.

## Industrial Wastewater Treatment

Industrial wastewater treatment covers the mechanisms and processes used to treat wastewater that is produced as a by-product of industrial or commercial activities. After treatment, the treated industrial wastewater (or effluent) may be reused or released to a sanitary sewer or to a surface water in the environment. Most industries produce some wastewater although recent trends in the developed world have been to minimise such production or recycle such wastewater within the production process. However, many industries remain dependent on processes that produce wastewaters.

### Sources of Industrial Wastewater

### Complex organic Chemicals Industry

A range of industries manufacture or use complex organic chemicals. These include pesticides, pharmaceuticals, paints and dyes, petrochemicals, detergents, plastics, paper pollution, etc. Waste waters can be contaminated by feedstock materials, by-products, product material in soluble or particulate form, washing and cleaning agents, solvents and added value products such as plasticisers. Treatment facilities that do not need control of their effluent typically opt for a type of aerobic treatment, i.e. aerated lagoons.

### Electric Power Plants

Fossil-fuel power stations, particularly coal-fired plants, are a major source of industrial wastewater. Many of these plants discharge wastewater with significant levels of metals such as lead, mercury, cadmium and chromium, as well as arsenic, selenium and nitrogen compounds (nitrates and nitrites). Wastewater streams include flue-gas desulfurization, fly ash, bottom ash and flue gas mercury control. Plants with air pollution controls such as wet scrubbers typically transfer the captured pollutants to the wastewater stream.

Ash ponds, a type of surface impoundment, are a widely used treatment technology at coal-fired plants. These ponds use gravity to settle out large particulates (measured as total suspended solids) from power plant wastewater. This technology does not treat dissolved pollutants. Power stations use additional technologies to control pollutants, depending on the particular wastestream in the plant. These include dry ash handling, closed-loop ash recycling, chemical precipitation, biological treatment (such as an activated sludge process), membrane systems, and evaporation-crystallization systems. Technological advancements in ion exchange membranes and electrodialysis systems has enabled high efficiency treatment of flue-gas desulfurization wastewater to meet recent EPA discharge limits. The treatment approach is similar for other highly scaling industrial wastewaters.

## Food Industry

Wastewater generated from agricultural and food operations has distinctive characteristics that set it apart from common municipal wastewater managed by public or private sewage treatment plants throughout the world: it is biodegradable and non-toxic, but has high concentrations of biochemical oxygen demand (BOD) and suspended solids (SS). The constituents of food and agriculture wastewater are often complex to predict, due to the differences in BOD and pH in effluents from vegetable, fruit, and meat products and due to the seasonal nature of food processing and post-harvesting.

Processing of food from raw materials requires large volumes of high grade water. Vegetable washing generates waters with high loads of particulate matter and some dissolved organic matter. It may also contain surfactants.

Animal slaughter and processing produces very strong organic waste from body fluids, such as blood, and gut contents. This wastewater is frequently contaminated by significant levels of antibiotics and growth hormones from the animals and by a variety of pesticides used to control external parasites.

Processing food for sale produces wastes generated from cooking which are often rich in plant organic material and may also contain salt, flavourings, colouring material and acids or alkali. Very significant quantities of oil or fats may also be present.

## Iron and Steel Industry

The production of iron from its ores involves powerful reduction reactions in blast furnaces. Cooling waters are inevitably contaminated with products especially ammonia and cyanide. Production of coke from coal in coking plants also requires water cooling and the use of water in by-products separation. Contamination of waste streams includes gasification products such as benzene, naphthalene, anthracene, cyanide, ammonia, phenols, cresols together with a range of more complex organic compounds known collectively as polycyclic aromatic hydrocarbons (PAH).

The conversion of iron or steel into sheet, wire or rods requires hot and cold mechanical transformation stages frequently employing water as a lubricant and coolant. Contaminants include hydraulic oils, tallow and particulate solids. Final treatment of iron and steel products before onward sale into manufacturing includes *pickling* in strong mineral acid to remove rust and prepare the surface for tin or chromium plating or for other surface treatments such as galvanisation or painting. The two acids commonly used are hydrochloric acid and sulfuric acid. Wastewaters include acidic rinse waters together with waste acid. Although many plants operate acid recovery plants (particularly those using hydrochloric acid), where the mineral acid is boiled away from the iron salts, there remains a large volume of highly acid ferrous sulfate or ferrous chloride to be disposed of. Many steel industry wastewaters are contaminated by hydraulic oil, also known as *soluble oil.*

## Mines and Quarries

Mine wastewater effluent in Peru, with neutralized pH from tailing runoff.

The principal waste-waters associated with mines and quarries are slurries of rock particles in water. These arise from rainfall washing exposed surfaces and haul roads and also from rock washing and grading processes. Volumes of water can be very high, especially rainfall related arisings on large sites. Some specialized separation operations, such as coal washing to separate coal from native rock using density gradients, can produce wastewater contaminated by fine particulate haematite and surfactants. Oils and hydraulic oils are also common contaminants.

Wastewater from metal mines and ore recovery plants are inevitably contaminated by the minerals present in the native rock formations. Following crushing and extraction of the desirable materials, undesirable materials may enter the wastewater stream. For metal mines, this can include unwanted metals such as zinc and other materials such as arsenic. Extraction of high value metals such as gold and silver may generate slimes containing very fine particles in where physical removal of contaminants becomes particularly difficult.

Additionally, the geologic formations that harbour economically valuable metals such as copper and gold very often consist of sulphide-type ores. The processing entails grinding the rock into fine particles and then extracting the desired metal(s), with the leftover rock being known as tailings. These tailings contain a combination of not only undesirable leftover metals, but also sulphide components which eventually form sulphuric acid upon the exposure to air and water that inevitably occurs when the tailings are disposed of in large impoundments. The resulting acid mine drainage, which is often rich in heavy metals (because acids dissolve metals), is one of the many environmental impacts of mining.

## Nuclear Industry

The waste production from the nuclear and radio-chemicals industry is dealt with as *Radioactive waste.*

## Pulp and Paper Industry

Effluent from the pulp and paper industry is generally high in suspended solids and BOD. Plants that bleach wood pulp for paper making may generate chloroform, dioxins (including 2,3,7,8-

TCDD), furans, phenols and chemical oxygen demand (COD). Stand-alone paper mills using imported pulp may only require simple primary treatment, such as sedimentation or dissolved air flotation. Increased BOD or COD loadings, as well as organic pollutants, may require biological treatment such as activated sludge or upflow anaerobic sludge blanket reactors. For mills with high inorganic loadings like salt, tertiary treatments may be required, either general membrane treatments like ultrafiltration or reverse osmosis or treatments to remove specific contaminants, such as nutrients.

## Industrial Oil Contamination

Industrial applications where oil enters the wastewater stream may include vehicle wash bays, workshops, fuel storage depots, transport hubs and power generation. Often the wastewater is discharged into local sewer or trade waste systems and must meet local environmental specifications. Typical contaminants can include solvents, detergents, grit. lubricants and hydrocarbons.

## Water Treatment

Many industries have a need to treat water to obtain very high quality water for demanding purposes such as environmental discharge compliance. Water treatment produces organic and mineral sludges from filtration and sedimentation. Ion exchange using natural or synthetic resins removes calcium, magnesium and carbonate ions from water, typically replacing them with sodium, chloride, hydroxyl and/or other ions. Regeneration of ion exchange columns with strong acids and alkalis produces a wastewater rich in hardness ions which are readily precipitated out, especially when in admixture with other wastewater constituents.

## Wool Processing

Insecticide residues in fleeces are a particular problem in treating waters generated in wool processing. Animal fats may be present in the wastewater, which if not contaminated, can be recovered for the production of tallow or further rendering.

## Treatment of Industrial Wastewater

The various types of contamination of wastewater require a variety of strategies to remove the contamination.

## Brine Treatment

Brine treatment involves removing dissolved salt ions from the waste stream. Although similarities to seawater or brackish water desalination exist, industrial brine treatment may contain unique combinations of dissolved ions, such as hardness ions or other metals, necessitating specific processes and equipment.

Brine treatment systems are typically optimized to either reduce the volume of the final discharge for more economic disposal (as disposal costs are often based on volume) or maximize the recovery of fresh water or salts. Brine treatment systems may also be optimized to reduce electricity consumption, chemical usage, or physical footprint.

Brine treatment is commonly encountered when treating cooling tower blowdown, produced water from steam assisted gravity drainage (SAGD), produced water from natural gas extraction such as coal seam gas, frac flowback water, acid mine or acid rock drainage, reverse osmosis reject, chlor-alkali wastewater, pulp and paper mill effluent, and waste streams from food and beverage processing.

Brine treatment technologies may include: membrane filtration processes, such as reverse osmosis; ion exchange processes such as electrodialysis or weak acid cation exchange; or evaporation processes, such as brine concentrators and crystallizers employing mechanical vapour recompression and steam.

Reverse osmosis may not be viable for brine treatment, due to the potential for fouling caused by hardness salts or organic contaminants, or damage to the reverse osmosis membranes from hydrocarbons.

Evaporation processes are the most widespread for brine treatment as they enable the highest degree of concentration, as high as solid salt. They also produce the highest purity effluent, even distillate-quality. Evaporation processes are also more tolerant of organics, hydrocarbons, or hardness salts. However, energy consumption is high and corrosion may be an issue as the prime mover is concentrated salt water. As a result, evaporation systems typically employ titanium or duplex stainless steel materials.

## Brine Management

Brine management examines the broader context of brine treatment and may include consideration of government policy and regulations, corporate sustainability, environmental impact, recycling, handling and transport, containment, centralized compared to on-site treatment, avoidance and reduction, technologies, and economics. Brine management shares some issues with leachate management and more general waste management.

## Solids Removal

Most solids can be removed using simple sedimentation techniques with the solids recovered as slurry or sludge. Very fine solids and solids with densities close to the density of water pose special problems. In such case filtration or ultrafiltration may be required. Although, flocculation may be used, using alum salts or the addition of polyelectrolytes.

## Oils and Grease Removal

The effective removal of oils and grease is dependent on the characteristics of the oil in terms of its suspension state and droplet size, which will in turn affect the choice of separator technology.

Oil pollution in water usually comes in four states, often in combination:

- free oil - large oil droplets sitting on the surface;
- heavy oil, which sits at the bottom, often adhering to solids like dirt;
- emulsified, where the oil droplets are heavily “chopped”; and

- dissolved oil, where the droplets are fully dispersed and not visible. Emulsified oil droplets are the most common in industrial oily wastewater and are extremely difficult to separate.

The methodology for separating the oil is dependent on the oil droplet size. Larger oil droplets such as those in free oil pollution are easily removed, but as the droplets become smaller, some separator technologies perform better than others.

Most separator technologies will have an optimum range of oil droplet sizes that can be effectively treated. This is known as the "micron rating."

Analysing the oily water to determine droplet size can be performed with a video particle analyser. Alternatively, there are commonalities in industries for oil droplet sizes. Larger droplets–greater than 60 microns–are often present in wastewater in workshops, re-fuel areas and depots. Twenty to 50 micron oil droplets often are present in vehicle wash bays, meat processing and dairy manufacturing effluent and aluminium billet cooling towers. Smaller droplets in the range of 10 to 20 microns tend to occur in workshops and condensates.

Each separator technology will have its' own performance curve outlining optimum performance based on oil droplet size. the most common separators are gravity tanks or pits, API oil-water separators or plate packs, chemical treatment via DAFs, centrifuges, media filters and hydrocyclones.

## API Separators

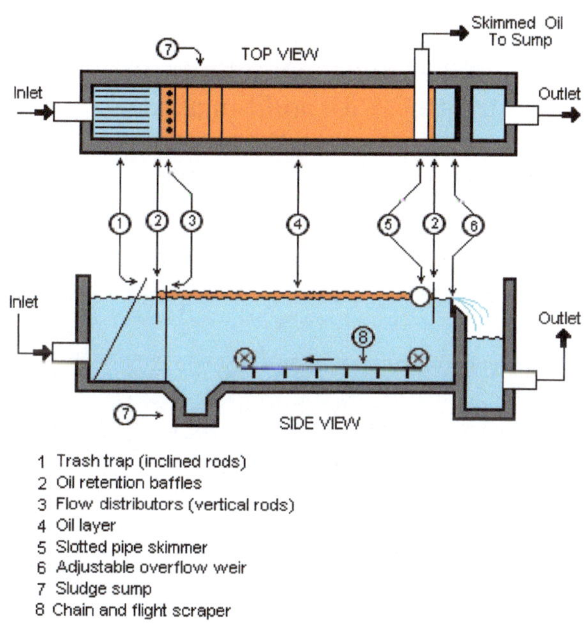


A typical API oil-water separator used in many industries

Many oils can be recovered from open water surfaces by skimming devices. Considered a dependable and cheap way to remove oil, grease and other hydrocarbons from water, oil skimmers can sometimes achieve the desired level of water purity. At other times, skimming is also a cost-efficient method to remove most of the oil before using membrane filters and chemical processes. Skimmers will prevent filters from blinding prematurely and keep chemical costs down because there is less oil to process.

Because grease skimming involves higher viscosity hydrocarbons, skimmers must be equipped with heaters powerful enough to keep grease fluid for discharge. If floating grease forms into solid clumps or mats, a spray bar, aerator or mechanical apparatus can be used to facilitate removal.

However, hydraulic oils and the majority of oils that have degraded to any extent will also have a soluble or emulsified component that will require further treatment to eliminate. Dissolving or emulsifying oil using surfactants or solvents usually exacerbates the problem rather than solving it, producing wastewater that is more difficult to treat.

The wastewaters from large-scale industries such as oil refineries, petrochemical plants, chemical plants, and natural gas processing plants commonly contain gross amounts of oil and suspended solids. Those industries use a device known as an API oil-water separator which is designed to separate the oil and suspended solids from their wastewater effluents. The name is derived from the fact that such separators are designed according to standards published by the American Petroleum Institute (API).

The API separator is a gravity separation device designed by using Stokes Law to define the rise velocity of oil droplets based on their density and size. The design is based on the specific gravity difference between the oil and the wastewater because that difference is much smaller than the specific gravity difference between the suspended solids and water. The suspended solids settles to the bottom of the separator as a sediment layer, the oil rises to top of the separator and the cleansed wastewater is the middle layer between the oil layer and the solids.

Typically, the oil layer is skimmed off and subsequently re-processed or disposed of, and the bottom sediment layer is removed by a chain and flight scraper (or similar device) and a sludge pump. The water layer is sent to further treatment for additional removal of any residual oil and then to some type of biological treatment unit for removal of undesirable dissolved chemical compounds.

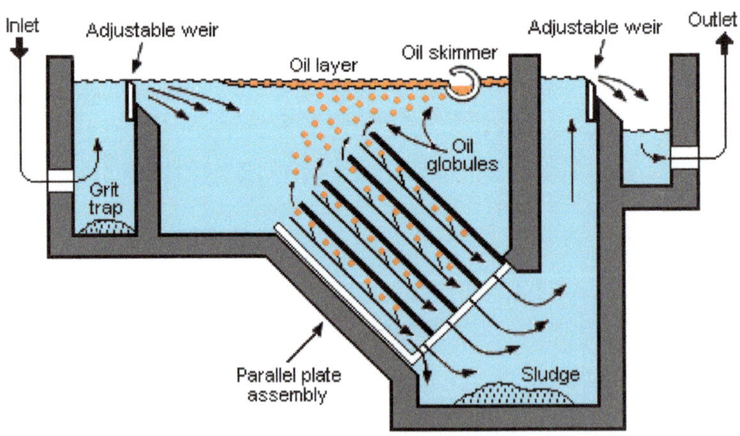


A typical parallel plate separator

Parallel plate separators are similar to API separators but they include tilted parallel plate assemblies (also known as parallel packs). The parallel plates provide more surface for suspended oil droplets to coalesce into larger globules. Such separators still depend upon the specific gravity between the suspended oil and the water. However, the parallel plates enhance the degree of oil-water separation. The result is that a parallel plate separator requires significantly less space than a conventional API separator to achieve the same degree of separation.

## Hydrocyclone Oil Separators

Hydrocyclone oil separators operate on the process where wastewater enters the cyclone chamber and is spun under extreme centrifugal forces more than 1000 times the force of gravity. This force causes the water and oil droplets to separate. The separated oil is discharged from one end of the cyclone where treated water is discharged through the opposite end for further treatment, filtration or discharge.

Hydrocyclones are useful for the greatest range of oil droplet sizes operating from less than 10 microns and up and can operate continuously without water pre-treatment and at any temperature and pH. Applications where hydrocyclones are found are in industry where oily water sources arise in workshops, vehicle wash bays, transport hubs, fuel depots and aluminium billet processing. Animal fats from meat processing and dairy manufacturing can also be removed without the need of chemical treatment that often is required for dissolved air flotation (DAF) systems.

## Removal of Biodegradable Organics

Biodegradable organic material of plant or animal origin is usually possible to treat using extended conventional sewage treatment processes such as activated sludge or trickling filter. Problems can arise if the wastewater is excessively diluted with washing water or is highly concentrated such as undiluted blood or milk. The presence of cleaning agents, disinfectants, pesticides, or antibiotics can have detrimental impacts on treatment processes.

## Activated Sludge Process

Activated sludge is a biochemical process for treating sewage and industrial wastewater that uses air (or oxygen) and microorganisms to biologically oxidize organic pollutants, producing a waste sludge (or floc) containing the oxidized material. In general, an activated sludge process includes:

- An aeration tank where air (or oxygen) is injected and thoroughly mixed into the wastewater.
- A settling tank (usually referred to as a clarifier or "settler") to allow the waste sludge to settle. Part of the waste sludge is recycled to the aeration tank and the remaining waste sludge is removed for further treatment and ultimate disposal.

## Trickling Filter Process

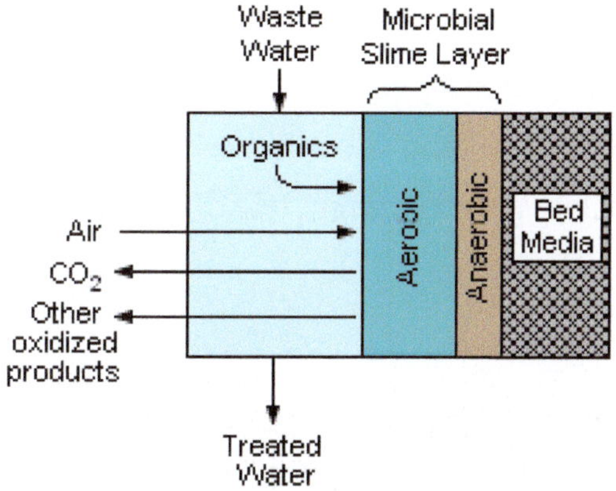


A schematic cross-section of the contact face of the bed media in a trickling filter

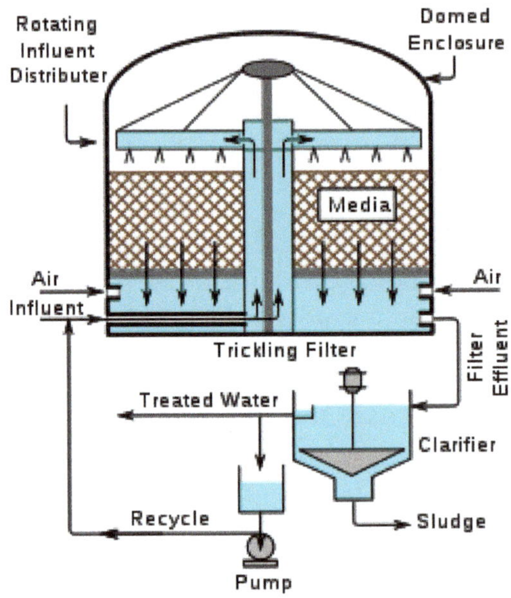


A typical complete trickling filter system

A trickling filter consists of a bed of rocks, gravel, slag, peat moss, or plastic media over which wastewater flows downward and contacts a layer (or film) of microbial slime covering the bed media. Aerobic conditions are maintained by forced air flowing through the bed or by natural convection of air. The process involves adsorption of organic compounds in the wastewater by the microbial slime layer, diffusion of air into the slime layer to provide the oxygen required for the biochemical oxidation of the organic compounds. The end products include carbon dioxide gas, water and other products of the oxidation. As the slime layer thickens, it becomes difficult for the air to penetrate the layer and an inner anaerobic layer is formed.

The fundamental components of a complete trickling filter system are:

- A bed of filter medium upon which a layer of microbial slime is promoted and developed.
- An enclosure or a container which houses the bed of filter medium.
- A system for distributing the flow of wastewater over the filter medium.
- A system for removing and disposing of any sludge from the treated effluent.

The treatment of sewage or other wastewater with trickling filters is among the oldest and most well characterized treatment technologies.

A trickling filter is also often called a *trickle filter, trickling biofilter, biofilter, biological filter* or *biological trickling filter.*

## Removal of Other Organics

Synthetic organic materials including solvents, paints, pharmaceuticals, pesticides, products from coke production and so forth can be very difficult to treat. Treatment methods are often specific to the material being treated. Methods include advanced oxidation processing, distillation, adsorption, vitrification, incineration, chemical immobilisation or landfill disposal. Some materials such as some detergents may be capable of biological degradation and in such cases, a modified form of wastewater treatment can be used.

## Removal of Acids and Alkalis

Acids and alkalis can usually be neutralised under controlled conditions. Neutralisation frequently produces a precipitate that will require treatment as a solid residue that may also be toxic. In some cases, gases may be evolved requiring treatment for the gas stream. Some other forms of treatment are usually required following neutralisation.

Waste streams rich in hardness ions as from de-ionisation processes can readily lose the hardness ions in a buildup of precipitated calcium and magnesium salts. This precipitation process can cause severe *furring* of pipes and can, in extreme cases, cause the blockage of disposal pipes. A 1 metre diameter industrial marine discharge pipe serving a major chemicals complex was blocked by such salts in the 1970s. Treatment is by concentration of de-ionisation waste waters and disposal to landfill or by careful pH management of the released wastewater.

## Removal of Toxic Materials

Toxic materials including many organic materials, metals (such as zinc, silver, cadmium, thallium, etc.) acids, alkalis, non-metallic elements (such as arsenic or selenium) are generally resistant to biological processes unless very dilute. Metals can often be precipitated out by changing the pH or by treatment with other chemicals. Many, however, are resistant to treatment or mitigation and may require concentration followed by landfilling or recycling. Dissolved organics can be *incinerated* within the wastewater by the advanced oxidation process.

## Smart Capsules

Molecular encapsulation is a technology that has the potential to provide a system for the recyclable removal of lead and other ions from polluted sources. Nano-, micro- and milli- capsules, with sizes in the ranges 10 nm-1μm,1μm-1mm and >1mm, respectively, are particles that have an active reagent (core) surrounded by a carrier (shell).There are three types of capsule under investigation: alginate-based capsules, carbon nanotubes, polymer swelling capsules. These capsules provide a possible means for the remediation of contaminated water.

## Powdered Activated Carbon Treatment

Powdered Activated Carbon Treatment (PACT) is a wastewater technology in which powdered activated carbon is added to an anaerobic or aerobic treatment system. The carbon in the biological treatment process acts as a "buffer" against the effects of toxic organics in the wastewater. Powdered Activated Carbon is also used in the processing of drinking water at treatment facilities, primarily on a seasonal basis in order to deal with aesthetic problems with the water such as odor and taste issues associated with Geosmin and 2-MIB.

In such a system, biological treatment and carbon adsorption are combined into a single, synergistic treatment step. The result is a system which offers significant cost reduction compared to activated sludge and granular carbon treatment options. The addition of the powdered activated carbon stabilizes biological systems against upsets and shock loading, controls color and odor, and may reduce disposal costs while removing soluble organics.

### Commercial Applications

The system as it exists today was developed in the 1970s under a collaborative effort between DuPont and Zimpro. PACT systems can be used to treat industrial wastewater, landfill leachate and highly contaminated surface water or groundwater.

In an aerobic treatment system, influent first enters an aeration tank where the powdered carbon is added, making up a portion of mixed liquor suspended solids. Once the aeration is completed, the treated wastewater and the carbon-biomass slurry are allowed to settle.

Following treatment, a portion of the carbon and biomass slurry is wasted to solids handling. These solids can be wasted and disposed of as a slurry, dewatered to a compact, stable cake, or pumped as a slurry to a wet air oxidation unit for further processing to regenerate the carbon and destroy the biological solids.

## Vapor Recovery

Vapor (or vapour) recovery is the process of recovering the vapors of gasoline or other fuels, so that they do not escape into the atmosphere. This is often done (or required by law) at filling stations, in order to reduce noxious and potentially explosive fumes and pollution.

The negative pressure created in the (underground) tank by the withdrawal is usually used to pull in the vapors. They are drawn-in through holes in the side of the nozzle and travel through special hoses which have a return path.

### State Requirements

- Illinois

### Other Industries

Vapor recovery is also used in the chemical process industry to remove and recover vapors from storage tanks. The vapors are usually either environmentally hazardous, or valuable to be recovered. The process consists of a closed venting system from the storage tank ullage space to a vapor recovery unit (VRU) which will recover the vapors for return to the process or destroy them, usually by oxidation.

Vapor recovery units are also becoming commonly used in the oil and gas industry as a means of recovering natural gas vapor and making it a usable and profitable product. Specifically a newer form of vapor recovery technology, Ejector Vapor Recovery Units create a closed loop system which not only recovers valuable vapor, but also reduces methane and VOC emissions.

In the Australian region vapor recovery has become mandatory in major urban areas. There are 2 categories - VR1 and VR2. VR1 is to be installed in fuel stations that pump less than 500,000 litres annually, VR2 is for fuel quantities over 500,000 litres per annum, or as designated by various EPA bodies.

Vapor recovery towers are also used in the oil and gas industry to provide flash gas recovery at near atmospheric pressure without the potential of oxygen ingress from the top of the storage tanks. The ability to create the vapor flash inside the vapor recovery tower often reduces storage tank emissions to less than 6 tons per year, exempting the tank battery from Quad O reporting requirements.

## References


- Ali Farnoud "Electrostatic Removal of Diesel Particulate Matter", ProQuest, 2008 ISBN 0549508163, p. 23
- Randall, Clifford W.; Sen, Dipankar (1996). "Full-scale evaluation of an integrated fixed-film activated sludge (IFAS) process for enhanced nitrogen removal". Water Science and Technology. Elsevier. 33 (12): 155–162. doi:10.1016/0273-1223(96)00469-6
- Caribbean Environment Programme (1998). Appropriate Technology for Sewage Pollution Control in the Wider Caribbean Region (PDF). Kingston, Jamaica: United Nations Environment Programme. Retrieved 2009-10-12
- Ashton, John; Ubido, Janet (1991). "The Healthy City and the Ecological Idea" (PDF). Journal of the Society for the Social History of Medicine. 4 (1): 173–181
- Khopkar, S. M. (2004). Environmental Pollution Monitoring And Control. New Delhi: New Age International. p. 299. ISBN 81-224-1507-5
- Margot, J.; et al. (2013). "Bacterial versus fungal laccase: potential for micropollutant degradation". AMB Express. 3: 63. doi:10.1186/2191-0855-3-63
- Environment Agency (archive) – Persistent, bioaccumulative and toxic PBT substances at the Wayback Machine (archived August 4, 2006). environment-agency.gov.uk. Retrieved on 2012-12-19
- Kashiwaya, Mamoru; Yoshimoto, Kameo (May 1980). "Tannery Wastewater Treatment by the Oxygen Activated Sludge Process". Journal (Water Pollution Control Federation). 52 (5): 999–1007. JSTOR 25040825
- Burton, Jr., G. Allen; Pitt, Robert E. (2001). "Chapter 2. Receiving Water Uses, Impairments, and Sources of Stormwater Pollutants". Stormwater Effects Handbook: A Toolbox for Watershed Managers, Scientists, and Engineers. New York: CRC/Lewis Publishers. ISBN 0-87371-924-7
- Caperon, J.; Cattell, S. A. & Krasnick, G. (1971). "Phytoplankton Kinetics in a Subtropical Estuary: Eutrophication" (PDF). Limnology and Oceanography. 16 (4): 599–607. doi:10.4319/lo.1971.16.4.0599
- Natural Environmental Research Council – River sewage pollution found to be disrupting fish hormones. Planetearth.nerc.ac.uk. Retrieved on 2012-12-19
- Haughey, A. (1968). "The Planktonic Algae of Auckland Sewage Treatment Ponds". New Zealand Journal of Marine and Freshwater Research. 2 (4): 721–766. doi:10.1080/00288330.1968.9515271
- Corcoran, E., C. Nellemann, E. Baker, R. Bos, D. Osborn, H. Savelli (eds) (2010). Sick water? : the central role of wastewater management in sustainable development : a rapid response assessment (PDF). Arendal, Norway: UNEP/GRID-Arendal. ISBN 978-82-7701-075-5
- Das, Tapas K. (August 2001). "Ultraviolet disinfection application to a wastewater treatment plant". Clean Technologies and Environmental Policy. Springer Berlin/Heidelberg. 3 (2): 69–80. doi:10.1007/S100980100108
- Griggs, Mary Beth (20 July 2015). "Mercury Scrubbers On Power Plants Clean Up Other Pollutants, Too". Popular Science. Retrieved 30 July 2015

- Benidickson, Jamie (2011). The Culture of Flushing: A Social and Legal History of Sewage. UBC Press. Retrieved 2013-02-07
- Tilley, David F. (2011). Aerobic Wastewater Treatment Processes: History and Development. IWA Publishing. ISBN 9781843395423
- Lienert, J.; Bürki, T.; Escher, B.I. (2007). "Reducing micropollutants with source control: Substance flow analysis of 212 pharmaceuticals in faeces and urine". Water Science & Technology. 56 (5): 87–96. doi:10.2166/wst.2007.560
- "Lowering Cost and Waste in Flue Gas Desulfurization Wastewater Treatment". Power Mag. Electric Power. Retrieved 6 April 2017
- Eckenfelder, Jr., W. Wesley; Cleary, Joseph G. (2014). Activated Sludge Technologies for Treating Industrial Wastewaters (1st ed.). DEStech Publications. p. 234. ISBN 978-1-60595-019-8. Retrieved 29 December 2014

4

# Pollution Prevention Practices

One of the practices used for preventing pollution is recycling. It is the reduction of waste by re-using products and materials and converting waste into useful products. The other methods of reducing pollution are waste minimization, pollution prevention, regulation and monitoring of pollution, etc. This chapter will help the readers in developing a better understanding of air, noise and water pollution.

## Pollution Prevention

Pollution prevention reduces the amount of pollution generated by a process, whether by industry, agriculture, or consumers. In contrast to most pollution control strategies,which seek to manage a pollutant after it is emitted and reduce its impact upon the environment, the pollution prevention approach seeks to increase the efficiency of a process, the reducing the amount of pollution generated at its source. Although there is wide agreement that source reduction is the preferred strategy, some professionals also use the term pollution prevention to include pollution reduction.

With increasing human population, pollution has become a great concern. Pollution from human activities is a problem that does not have to be inevitable. With a comprehensive pollution prevention program, most pollution can be reduced, reused, or prevented. The US Environmental Protection Agency works to introduce pollution prevention programs to reduce and manage waste. Reducing and managing pollution may decrease the number of deaths and illnesses from pollution-related diseases.

### Human Population Numbers

With nearly seven billion people in the world, and the fact that the average person produces 4.4 pounds of waste each day, around a ton of waste every year, it is easy to see why pollution is such a huge problem (Recycling Revolution, 2010). One way to slow the growing levels of pollution would be to stabilize the human population. Population numbers are rapidly increasing in developing countries. It is estimated that "the human population will increase by one billion people in the next decade" (Kinder, 2011, Sec. 1, Para. 1). Growth in these developing countries is partly due to their governments telling their citizens that more numbers are needed to fill in the open spaces as the West has (Kinder, 2011).

Some advocates think that people should have fewer children. "Experience shows that the most effective ways to slow human population growth are to encourage family planning & to reduce poverty, and to elevate the status of women." (Miller & Spoolman, 2009, p. 133). Such plans and strategies can be converted into policies to ensure sustainability. "Action plans and strategies can be developed to increase public understanding of how rapid population growth limits chances for meeting basic... recycling or reuse."

## Production Techniques

As an environmental management strategy, pollution prevention shares many attributes with cleaner production, a term used more commonly outside the United States. Pollution prevention encompasses more specialized sub-disciplines including green chemistry and green design (also known as environmentally conscious design).

# Recycling

The three chasing arrows of the international recycling logo.
It is sometimes accompanied by the text "reduce, reuse, and recycle".

Recycling is the process of converting waste materials into new materials and objects. It is an alternative to "conventional" waste disposal that can save material and help lower greenhouse gas emissions (compared to plastic production, for example). Recycling can prevent the waste of potentially useful materials and reduce the consumption of fresh raw materials, thereby reducing: energy usage, air pollution (from incineration), and water pollution (from landfilling).

Recycling is a key component of modern waste reduction and is the third component of the "Reduce, Reuse, and Recycle" waste hierarchy.

There are some ISO standards related to recycling such as ISO 15270:2008 for plastics waste and ISO 14001:2004 for environmental management control of recycling practice.

*Recyclable materials* include many kinds of glass, paper, and cardboard, metal, plastic, tires, textiles, and electronics. The composting or other reuse of biodegradable waste—such as food or garden waste—is also considered recycling. Materials to be recycled are either brought to a collection centre or picked up from the curbside, then sorted, cleaned, and reprocessed into new materials destined for manufacturing.

In the strictest sense, recycling of a material would produce a fresh supply of the same material—for example, used office paper would be converted into new office paper or used polystyrene foam into new polystyrene. However, this is often difficult or too expensive (compared with producing the same product from raw materials or other sources), so "recycling" of many products or materials involves their *reuse* in producing different materials (for example, paperboard) instead.

Another form of recycling is the salvage of certain materials from complex products, either due to their intrinsic value (such as lead from car batteries, or gold from circuit boards), or due to their hazardous nature (e.g., removal and reuse of mercury from thermometers and thermostats).

## History

### Origins

Recycling has been a common practice for most of human history, with recorded advocates as far back as Plato in 400 BC. During periods when resources were scarce and hard to come by, archaeological studies of ancient waste dumps show less household waste (such as ash, broken tools, and pottery)—implying more waste was being recycled in the absence of new material.

An American poster from World War II

In pre-industrial times, there is evidence of scrap bronze and other metals being collected in Europe and melted down for perpetual reuse. Paper recycling was first recorded in 1031 when Japanese shops sold repulped paper. In Britain dust and ash from wood and coal fires was collected by "dustmen" and downcycled as a base material used in brick making. The main driver for these types of recycling was the economic advantage of obtaining recycled feedstock instead of acquiring virgin material, as well as a lack of public waste removal in ever more densely populated areas. In 1813, Benjamin Law developed the process of turning rags into "shoddy" and "mungo" wool in Batley, Yorkshire. This material combined recycled fibers with virgin wool. The West Yorkshire shoddy industry in towns such as Batley and Dewsbury lasted from the early 19th century to at least 1914.

Industrialization spurred demand for affordable materials; aside from rags, ferrous scrap metals were coveted as they were cheaper to acquire than virgin ore. Railroads both purchased and sold scrap metal in the 19th century, and the growing steel and automobile industries purchased scrap in the early 20th century. Many secondary goods were collected, processed and sold by peddlers who scoured dumps and city streets for discarded machinery, pots, pans, and other sources of metal. By World War I, thousands of such peddlers roamed the streets of American cities, taking advantage of market forces to recycle post-consumer materials back into industrial production.

Beverage bottles were recycled with a refundable deposit at some drink manufacturers in Great Britain and Ireland around 1800, notably Schweppes. An official recycling system with refundable deposits was established in Sweden for bottles in 1884 and aluminum beverage cans in 1982; the law led to a recycling rate for beverage containers of 84–99 percent depending on type, and a glass bottle can be refilled over 20 times on average.

British poster from World War II

## Wartime

New chemical industries created in the late 19th century both invented new materials (e.g. Bakelite (1907) and promised to transform valueless into valuable materials. Proverbially, you could not make a silk purse of a sow's ear—until the US firm Arhur D. Little published in 1921 "On the Making of Silk Purses from Sows' Ears", its research proving that when "chemistry puts on overalls and gets down to business . . .new values appear. New and better paths are opened to reach the goals desired."

Recycling was a highlight throughout World War II. During the war, financial constraints and significant material shortages due to war efforts made it necessary for countries to reuse goods and recycle materials. These resource shortages caused by the world wars, and other such world-changing occurrences, greatly encouraged recycling. The struggles of war claimed much of the material resources available, leaving little for the civilian population. It became necessary for most homes to recycle their waste, as recycling offered an extra source of materials allowing people to make the most of what was available to them. Recycling household materials meant more resources for war efforts and a better chance of victory. Massive government promotion campaigns were carried out in the home front during World War II in every country involved in the war, urging citizens to donate metals and conserve fiber, as a matter of patriotism.

## Post-war

A considerable investment in recycling occurred in the 1970s, due to rising energy costs. Recycling aluminum uses only 5% of the energy required by virgin production; glass, paper and other metals have less dramatic but very significant energy savings when recycled feedstock is used.

Although consumer electronics such as the television have been popular since the 1920s, recycling of them was almost unheard of until early 1991. The first electronic waste recycling scheme was implemented in Switzerland, beginning with collection of old refrigerators but gradually expanding to cover all devices. After these schemes were set up, many countries did not have the capacity to deal with the sheer quantity of e-waste they generated or its hazardous nature. They began to export the problem to developing countries without enforced environmental legislation. This is cheaper, as recycling computer monitors in the United States costs 10 times more than in China. Demand in Asia for electronic waste began to grow when scrap yards found that they could extract valuable substances such as copper, silver, iron, silicon, nickel, and gold, during the recycling process. The 2000s saw a large increase in both the sale of electronic devices and their growth as a waste stream: in 2002, e-waste grew faster than any other type of waste in the EU. This caused investment in modern, automated facilities to cope with the influx of redundant appliances, especially after strict laws were implemented in 2003.

As of 2014, the European Union has about 50% of world share of the waste and recycling industries, with over 60,000 companies employing 500,000 persons, with a turnover of €24 billion. Countries have to reach recycling rates of at least 50%, while the lead countries are around 65% and the EU average is 39% as of 2013.

## Legislation

### Supply

For a recycling program to work, having a large, stable supply of recyclable material is crucial. Three legislative options have been used to create such a supply: mandatory recycling collection, container deposit legislation, and refuse bans. Mandatory collection laws set recycling targets for cities to aim for, usually in the form that a certain percentage of a material must be diverted from the city's waste stream by a target date. The city is then responsible for working to meet this target.

Container deposit legislation involves offering a refund for the return of certain containers, typically glass, plastic, and metal. When a product in such a container is purchased, a small surcharge is added to the price. This surcharge can be reclaimed by the consumer if the container is returned to a collection point. These programs have been very successful, often resulting in an 80 percent recycling rate. Despite such good results, the shift in collection costs from local government to industry and consumers has created strong opposition to the creation of such programs in some areas. A variation on this is where the manufacturer bears responsibility for the recycling of their goods. In the European Union, the WEEE Directive requires producers of consumer electronics to reimburse the recyclers' costs.

An alternative way to increase supply of recyclates is to ban the disposal of certain materials as waste, often including used oil, old batteries, tires, and garden waste. One aim of this method is to create a viable economy for proper disposal of banned products. Care must be taken that enough of these recycling services exist, or such bans simply lead to increased illegal dumping.

## Government-mandated Demand

Legislation has also been used to increase and maintain a demand for recycled materials. Four methods of such legislation exist: minimum recycled content mandates, utilization rates, procurement policies, and recycled product labeling.

Both minimum recycled content mandates and utilization rates increase demand directly by forcing manufacturers to include recycling in their operations. Content mandates specify that a certain percentage of a new product must consist of recycled material. Utilization rates are a more flexible option: industries are permitted to meet the recycling targets at any point of their operation or even contract recycling out in exchange for tradeable credits. Opponents to both of these methods point to the large increase in reporting requirements they impose, and claim that they rob industry of necessary flexibility.

Governments have used their own purchasing power to increase recycling demand through what are called "procurement policies." These policies are either "set-asides," which reserve a certain amount of spending solely towards recycled products, or "price preference" programs which provide a larger budget when recycled items are purchased. Additional regulations can target specific cases: in the United States, for example, the Environmental Protection Agency mandates the purchase of oil, paper, tires and building insulation from recycled or re-refined sources whenever possible.

The final government regulation towards increased demand is recycled product labeling. When producers are required to label their packaging with amount of recycled material in the product (including the packaging), consumers are better able to make educated choices. Consumers with sufficient buying power can then choose more environmentally conscious options, prompt producers to increase the amount of recycled material in their products, and indirectly increase demand. Standardized recycling labeling can also have a positive effect on supply of recyclates if the labeling includes information on how and where the product can be recycled.

## Recyclates

Recyclate is a raw material that is sent to, and processed in a waste recycling plant or materials recovery facility which will be used to form new products. The material is collected in various methods and delivered to a facility where it undergoes re-manufacturing so that it can be used in the production of new materials or products. For example, plastic bottles that are collected can be re-used and made into plastic pellets, a new product.

## Quality of Recyclate

The quality of recyclates is recognized as one of the principal challenges that needs to be addressed for the success of a long-term vision of a green economy and achieving zero waste. Recyclate quality is generally referring to how much of the raw material is made up of target material compared to the amount of non-target material and other non-recyclable material. Only target material is likely to be recycled, so a higher amount of non-target and non-recyclable material will reduce the quantity of recycling product. A high proportion of non-target and non-recyclable material can make it more difficult for re-processors to achieve "high-quality" recycling. If the recyclate is of poor quality, it is more likely to end up being down-cycled or, in more extreme cases, sent to other recovery options or landfilled. For example, to facilitate the re-manufacturing of clear glass products there are tight restrictions for colored glass going into the re-melt process.

The quality of recyclate not only supports high-quality recycling, but it can also deliver significant environmental benefits by reducing, reusing and keeping products out of landfills. High-quality recycling can help support growth in the economy by maximizing the economic value of the waste material collected. Higher income levels from the sale of quality recyclates can return value which can be significant to local governments, households, and businesses. Pursuing high-quality recycling can also provide consumer and business confidence in the waste and resource management sector and may encourage investment in that sector.

There are many actions along the recycling supply chain that can influence and affect the material quality of recyclate. It begins with the waste producers who place non-target and non-recyclable wastes in recycling collection. This can affect the quality of final recyclate streams or require further efforts to discard those materials at later stages in the recycling process. The different collection systems can result in different levels of contamination. Depending on which materials are collected together, extra effort is required to sort this material back into separate streams and can significantly reduce the quality of the final product. Transportation and the compaction of materials can make it more difficult to separate material back into separate waste streams. Sorting facilities are not one hundred per cent effective in separating materials, despite improvements in technology and quality recyclate which can see a loss in recyclate quality. The storage of materials outside where the product can become wet can cause problems for re-processors. Reprocessing facilities may require further sorting steps to further reduce the amount of non-target and non-recyclable material. Each action along the recycling path plays a part in the quality of recyclate.

## Quality Recyclate Action Plan (Scotland)

The Recyclate Quality Action Plan of Scotland sets out a number of proposed actions that the Scottish Government would like to take forward in order to drive up the quality of the materials being collected for recycling and sorted at materials recovery facilities before being exported or sold on to the reprocessing market.

The plan's objectives are to:

- Drive up the quality of recyclate.
- Deliver greater transparency about the quality of recyclate.
- Provide help to those contracting with materials recycling facilities to identify what is required of them
- Ensure compliance with the Waste (Scotland) regulations 2012.

- Stimulate a household market for quality recyclate.
- Address and reduce issues surrounding the Waste Shipment Regulations.

The plan focuses on three key areas, with fourteen actions which were identified to increase the quality of materials collected, sorted and presented to the processing market in Scotland.

The three areas of focus are:

1. Collection systems and input contamination
2. Sorting facilities – material sampling and transparency
3. Material quality benchmarking and standards

## Recycling Consumer Waste

### Collection

A three-sided bin at a railway station in Germany, intended to separate paper *(left)* and plastic wrappings *(right)* from other waste *(back)*

A number of different systems have been implemented to collect recyclates from the general waste stream. These systems lie along the spectrum of trade-off between public convenience and government ease and expense. The three main categories of collection are "drop-off centers," "buy-back centers", and "curbside collection."

### Curbside Collection

Curbside collection encompasses many subtly different systems, which differ mostly on where in the process the recyclates are sorted and cleaned. The main categories are mixed waste collection, commingled recyclables, and source separation. A waste collection vehicle generally picks up the waste.

A recycling truck collecting the contents of a recycling bin in Canberra, Australia

At one end of the spectrum is mixed waste collection, in which all recyclates are collected mixed in with the rest of the waste, and the desired material is then sorted out and cleaned at a central sorting facility. This results in a large amount of recyclable waste, paper especially, being too soiled to reprocess, but has advantages as well: the city need not pay for a separate collection of recyclates and no public education is needed. Any changes to which materials are recyclable is easy to accommodate as all sorting happens in a central location.

In a commingled or single-stream system, all recyclables for collection are mixed but kept separate from other waste. This greatly reduces the need for post-collection cleaning but does require public education on what materials are recyclable.

Source separation is the other extreme, where each material is cleaned and sorted prior to collection. This method requires the least post-collection sorting and produces the purest recyclates, but incurs additional operating costs for collection of each separate material. An extensive public education program is also required, which must be successful if recyclate contamination is to be avoided.

Source separation used to be the preferred method due to the high sorting costs incurred by commingled (mixed waste) collection. Advances in sorting technology, however, have lowered this overhead substantially—many areas which had developed source separation programs have since switched to co-mingled collection.

## Buy-back Centers

Buy-back centers differ in that the cleaned recyclates are purchased, thus providing a clear incentive for use and creating a stable supply. The post-processed material can then be sold. If this is profitable, this conserves the emission of greenhouse gases; if unprofitable, it increases the emission of greenhouse gasses. Government subsidies are necessary to make buy-back centres a viable enterprise. In 1993, according to the U.S. National Waste & Recycling Association, it costs on average US$50 to process a ton of material, which can be resold for US$30.

In the US, the value per ton of mixed recyclables was US$180 in 2011, US$80 in 2015, and $US$100 in 2017.

In 2017, glass is essentially valueless, because of the low cost of sand, its major component; low oil costs thwarts plastic recycling.

In 2017, Napa, California was reimbursed about 20% of its costs in recycling.

## Drop-off Centers

Drop-off centers require the waste producer to carry the recyclates to a central location, either an installed or mobile collection station or the reprocessing plant itself. They are the easiest type of collection to establish but suffer from low and unpredictable throughput.

## Distributed Recycling

For some waste materials such as plastic, recent technical devices called recyclebots enable a form of distributed recycling. Preliminary life-cycle analysis (LCA) indicates that such distributed recycling of HDPE to make filament of 3-D printers in rural regions is energetically favorable to either using virgin resin or conventional recycling processes because of reductions in transportation energy.

## Sorting

Once commingled recyclates are collected and delivered to a central collection facility, the different types of materials must be sorted. This is done in a series of stages, many of which involve automated processes such that a truckload of material can be fully sorted in less than an hour. Some plants can now sort the materials automatically, known as single-stream recycling. In plants, a variety of materials is sorted such as paper, different types of plastics, glass, metals, food scraps, and most types of batteries. A 30 percent increase in recycling rates has been seen in the areas where these plants exist.

Initially, the commingled recyclates are removed from the collection vehicle and placed on a conveyor belt spread out in a single layer. Large pieces of corrugated fiberboard and plastic bags are removed by hand at this stage, as they can cause later machinery to jam.

Early sorting of recyclable materials: glass and plastic bottles in Poland

Next, automated machinery such as disk screens and air classifiers separate the recyclates by weight, splitting lighter paper and plastic from heavier glass and metal. Cardboard is removed from the mixed paper and the most common types of plastic, PET (#1) and HDPE (#2), are collected. This separation is usually done by hand but has become automated in some sorting centers: a spectroscopic scanner is used to differentiate between different types of paper and plastic based on the absorbed wavelengths, and subsequently divert each material into the proper collection channel.

Strong magnets are used to separate out ferrous metals, such as iron, steel, and tin cans. Non-ferrous metals are ejected by magnetic eddy currents in which a rotating magnetic field induces an electric current around the aluminum cans, which in turn creates a magnetic eddy current inside the cans. This magnetic eddy current is repulsed by a large magnetic field, and the cans are ejected from the rest of the recyclate stream.

A recycling point in New Byth, Scotland, with separate containers for paper, plastics, and differently colored glass

Finally, glass is sorted according to its color: brown, amber, green, or clear. It may either be sorted by hand, or via an automated machine that uses colored filters to detect different colors. Glass fragments smaller than 10 millimetres (0.39 in) across cannot be sorted automatically, and are mixed together as "glass fines."

This process of recycling as well as reusing the recycled material has proven advantageous because it reduces amount of waste sent to landfills, conserves natural resources, saves energy, reduces greenhouse gas emissions, and helps create new jobs. Recycled materials can also be converted into new products that can be consumed again, such as paper, plastic, and glass.

The City and County of San Francisco's Department of the Environment is attempting to achieve a citywide goal of generating zero waste by 2020. San Francisco's refuse hauler, Recology, operates an effective recyclables sorting facility in San Francisco, which helped San Francisco reach a record-breaking diversion rate of 80%.

## Rinsing

Food packaging should no longer contain any organic matter (organic matter, if any, needs to be placed in a biodegradable waste bin or be buried in a garden). Since no trace of biodegradable material is best kept in the packaging before placing it in a trash bag, some packaging also needs to be rinsed.

## Recycling Industrial Waste

Mounds of shredded rubber tires are ready for processing

Although many government programs are concentrated on recycling at home, a 64% of waste in the United Kingdom is generated by industry. The focus of many recycling programs done by industry is the cost–effectiveness of recycling. The ubiquitous nature of cardboard packaging makes cardboard a commonly recycled waste product by companies that deal heavily in packaged goods, like retail stores, warehouses, and distributors of goods. Other industries deal in niche or specialized products, depending on the nature of the waste materials that are present.

The glass, lumber, wood pulp and paper manufacturers all deal directly in commonly recycled materials; however, old rubber tires may be collected and recycled by independent tire dealers for a profit.

Levels of metals recycling are generally low. In 2010, the International Resource Panel, hosted by the United Nations Environment Programme (UNEP) published reports on metal stocks that exist within society and their recycling rates. The Panel reported that the increase in the use of metals during the 20th and into the 21st century has led to a substantial shift in metal stocks from below ground to use in applications within society above ground. For example, the in-use stock of copper in the USA grew from 73 to 238 kg per capita between 1932 and 1999.

The report authors observed that, as metals are inherently recyclable, the metal stocks in society can serve as huge mines above ground (the term "urban mining" has been coined with this idea in mind). However, they found that the recycling rates of many metals are very low. The report warned that the recycling rates of some rare metals used in applications such as mobile phones, battery packs for hybrid cars and fuel cells, are so low that unless future end-of-life recycling rates are dramatically stepped up these critical metals will become unavailable for use in modern technology.

The military recycles some metals. The U.S. Navy's Ship Disposal Program uses ship breaking to reclaim the steel of old vessels. Ships may also be sunk to create an artificial reef. Uranium is a very dense metal that has qualities superior to lead and titanium for many military and industrial uses. The uranium left over from processing it into nuclear weapons and fuel for nuclear reactors is called depleted uranium, and it is used by all branches of the U.S. military use for armour-piercing shells and shielding.

The construction industry may recycle concrete and old road surface pavement, selling their waste materials for profit.

Some industries, like the renewable energy industry and solar photovoltaic technology, in particular, are being proactive in setting up recycling policies even before there is considerable volume to their waste streams, anticipating future demand during their rapid growth.

Recycling of plastics is more difficult, as most programs are not able to reach the necessary level of quality. Recycling of PVC often results in downcycling of the material, which means only products of lower quality standard can be made with the recycled material. A new approach which allows an equal level of quality is the Vinyloop process. It was used after the London Olympics 2012 to fulfill the PVC Policy.

## E-waste Recycling

Microprocessors retrieved from waste stream

E-waste is a growing problem, accounting for 20-50 million metric tons of global waste per year according to the EPA. It is also the fastest growing waste stream in the EU. Many recyclers do not recycle e-waste responsibly. After the cargo barge Khian Sea dumped 14,000 metric tons of toxic ash in Haiti, the Basel Convention was formed to stem the flow of hazardous substances into poorer countries. They created the e-Stewards certification to ensure that recyclers are held to the highest standards for environmental responsibility and to help consumers identify responsible recyclers. This works alongside other prominent legislation, such as the Waste Electrical and Electronic Equipment Directive of the EU the United States National Computer Recycling Act, to prevent poisonous chemicals from entering waterways and the atmosphere.

In the recycling process, television sets, monitors, cell phones, and computers are typically tested for reuse and repaired. If broken, they may be disassembled for parts still having high value if labor is cheap enough. Other e-waste is shredded to pieces roughly 10 centimetres (3.9 in) in size, and manually checked to separate out toxic batteries and capacitors which contain poisonous metals. The remaining pieces are further shredded to 10 millimetres (0.39 in) particles and passed under a magnet to remove ferrous metals. An eddy current ejects non-ferrous metals, which are sorted by density either by a centrifuge or vibrating plates. Precious metals can be dissolved in acid, sorted, and smelted into ingots. The remaining glass and plastic fractions are separated by density and sold to re-processors. Television sets and monitors must be manually disassembled to remove lead from CRTs or the mercury backlight from LCDs.

## Plastic Recycling

A container for recycling used plastic spoons into material for 3D printing

Plastic recycling is the process of recovering scrap or waste plastic and reprocessing the material into useful products, sometimes completely different in form from their original state. For instance, this could mean melting down soft drink bottles and then casting them as plastic chairs and tables.

### Physical Recycling

Some plastics are remelted to form new plastic objects; for example, PET water bottles can be converted into polyester destined for clothing. A disadvantage of this type of recycling is that the molecular weight of the polymer can change further and the levels of unwanted substances in the plastic can increase with each remelt.

### Chemical Recycling

For some polymers, it is possible to convert them back into monomers, for example, PET can be treated with an alcohol and a catalyst to form a dialkyl terephthalate. The terephthalate diester can be used with ethylene glycol to form a new polyester polymer, thus making it possible to use the pure polymer again.

### Waste Plastic Pyrolysis to Fuel Oil

Another process involves conversion of assorted polymers into petroleum by a much less precise thermal depolymerization process. Such a process would be able to accept almost any polymer or mix of polymers, including thermoset materials such as vulcanized rubber tires and the biopolymers in feathers and other agricultural waste. Like natural petroleum, the chemicals produced can be used as fuels or as feedstock. A RESEM Technology plant of this type in Carthage, Missouri, USA, uses turkey waste as input material. Gasification is a similar process but is not technically recycling since polymers are not likely to become the result. Plastic Pyrolysis can convert petroleum based waste streams such as plastics into quality fuels, carbons. Given below is the list of suitable plastic raw materials for pyrolysis:

- Mixed plastic (HDPE, LDPE, PE, PP, Nylon, Teflon, PS, ABS, FRP, etc.)
- Mixed waste plastic from waste paper mill
- Multi-layered plastic

## Recycling Codes

Recycling codes on products

In order to meet recyclers' needs while providing manufacturers a consistent, uniform system, a coding system was developed. The recycling code for plastics was introduced in 1988 by the plastics industry through the Society of the Plastics Industry. Because municipal recycling programs traditionally have targeted packaging—primarily bottles and containers—the resin coding system offered a means of identifying the resin content of bottles and containers commonly found in the residential waste stream.

Plastic products are printed with numbers 1–7 depending on the type of resin. Type 1 (polyethylene terephthalate) is commonly found in soft drink and water bottles. Type 2 (high-density polyethylene) is found in most hard plastics such as milk jugs, laundry detergent bottles, and some dishware. Type 3 (polyvinyl chloride) includes items such as shampoo bottles, shower curtains, hula hoops, credit cards, wire jacketing, medical equipment, siding, and piping. Type 4 (low-density polyethylene) is found in shopping bags, squeezable bottles, tote bags, clothing, furniture, and carpet. Type 5 is polypropylene and makes up syrup bottles, straws, Tupperware, and some automotive parts. Type 6 is polystyrene and makes up meat trays, egg cartons, clamshell containers, and compact disc cases. Type 7 includes all other plastics such as bulletproof materials, 3- and 5-gallon water bottles, and sunglasses. Having a recycling code or the chasing arrows logo on a material is not an automatic indicator that a material is recyclable but rather an explanation of what the material is. Types 1 and 2 are the most commonly recycled.

## Economic Impact

Critics dispute the net economic and environmental benefits of recycling over its costs, and suggest that proponents of recycling often make matters worse and suffer from confirmation bias. Specifically, critics argue that the costs and energy used in collection and transportation detract from (and outweigh) the costs and energy saved in the production process; also that the jobs produced by the recycling industry can be a poor trade for the jobs lost in logging, mining, and other industries associated with production; and that materials such as paper pulp can only be recycled a few times before material degradation prevents further recycling.

The National Waste and Recycling Association (NWRA), reported in May 2015, that recycling and waste made a $6.7 billion economic impact in Ohio, U.S., and employed 14,000 people.

## Cost–benefit Analysis

| Environmental effects of recycling | | |
|---|---|---|
| **Material** | **Energy savings** | **Air pollution savings** |
| Aluminium | 95% | 95% |
| Cardboard | 24% | — |
| Glass | 5–30% | 20% |
| Paper | 40% | 73% |
| Plastics | 70% | — |
| Steel | 60% | — |

There is some debate over whether recycling is economically efficient. It is said that dumping 10,000 tons of waste in a landfill creates six jobs while recycling 10,000 tons of waste can create over 36 jobs. However, the cost effectiveness of creating the additional jobs remains unproven. According to the U.S. Recycling Economic Informational Study, there are over 50,000 recycling establishments that have created over a million jobs in the US. Two years after New York City declared that implementing recycling programs would be "a drain on the city," New York City leaders realized that an efficient recycling system could save the city over $20 million. Municipalities often see fiscal benefits from implementing recycling programs, largely due to the reduced landfill costs. A study conducted by the Technical University of Denmark according to the Economist found that in 83 percent of cases, recycling is the most efficient method to dispose of household waste. However, a 2004 assessment by the Danish Environmental Assessment Institute concluded that incineration was the most effective method for disposing of drink containers, even aluminium ones.

Fiscal efficiency is separate from economic efficiency. Economic analysis of recycling does not include what economists call externalities, which are unpriced costs and benefits that accrue to individuals outside of private transactions. Examples include: decreased air pollution and greenhouse gases from incineration, reduced hazardous waste leaching from landfills, reduced energy consumption, and reduced waste and resource consumption, which leads to a reduction in environmentally damaging mining and timber activity. About 4,000 minerals are known, of these only a few hundred minerals in the world are relatively common. Known reserves of phosphorus will be exhausted within the next 100 years at current rates of usage. Without mechanisms such as taxes or subsidies to internalize externalities, businesses may ignore them despite the costs imposed on society. To make such nonfiscal benefits economically relevant, advocates have pushed for legislative action to increase the demand for recycled materials. The United States Environmental Protection Agency (EPA) has concluded in favor of recycling, saying that recycling efforts reduced the country's carbon emissions by a net 49 million metric tonnes in 2005. In the United Kingdom, the Waste and Resources Action Programme stated that Great Britain's recycling efforts reduce $CO_2$ emissions by 10–15 million tonnes a year. Recycling is more efficient in densely populated areas, as there are economies of scale involved.

Wrecked automobiles gathered for smelting

Certain requirements must be met for recycling to be economically feasible and environmentally effective. These include an adequate source of recyclates, a system to extract those recyclates from the waste stream, a nearby factory capable of reprocessing the recyclates, and a potential demand for the recycled products. These last two requirements are often overlooked—without both an industrial market for production using the collected materials and a consumer market for the manufactured goods, recycling is incomplete and in fact only "collection".

Free-market economist Julian Simon remarked "There are three ways society can organize waste disposal: (a) commanding, (b) guiding by tax and subsidy, and (c) leaving it to the individual and the market". These principles appear to divide economic thinkers today.

Frank Ackerman favours a high level of government intervention to provide recycling services. He believes that recycling's benefit cannot be effectively quantified by traditional *laissez-faire* economics. Allen Hershkowitz supports intervention, saying that it is a public service equal to education and policing. He argues that manufacturers should shoulder more of the burden of waste disposal.

Paul Calcott and Margaret Walls advocate the second option. A deposit refund scheme and a small refuse charge would encourage recycling but not at the expense of fly-tipping. Thomas C. Kinnaman concludes that a landfill tax would force consumers, companies and councils to recycle more.

Most free-market thinkers detest subsidy and intervention because they waste resources. Terry Anderson and Donald Leal think that all recycling programmes should be privately operated, and therefore would only operate if the money saved by recycling exceeds its costs. Daniel K. Benjamin argues that it wastes people's resources and lowers the wealth of a population.

## Trade in Recyclates

Certain countries trade in unprocessed recyclates. Some have complained that the ultimate fate of recyclates sold to another country is unknown and they may end up in landfills instead of reprocessed. According to one report, in America, 50–80 percent of computers destined for recycling are actually not recycled. There are reports of illegal-waste imports to China being dismantled and recycled solely for monetary gain, without consideration for workers' health or environmental damage. Although the Chinese government has banned these practices, it has not been able to eradicate them. In 2008, the prices of recyclable waste plummeted before rebounding in 2009. Cardboard averaged about £53/tonne from 2004–2008, dropped to £19/tonne, and then went up to £59/tonne in May 2009. PET plastic averaged about £156/tonne, dropped to £75/tonne and then moved up to £195/tonne in May 2009.

Certain regions have difficulty using or exporting as much of a material as they recycle. This problem is most prevalent with glass: both Britain and the U.S. import large quantities of wine bottled in green glass. Though much of this glass is sent to be recycled, outside the American Midwest there is not enough wine production to use all of the reprocessed material. The extra must be downcycled into building materials or re-inserted into the regular waste stream.

Similarly, the northwestern United States has difficulty finding markets for recycled newspaper, given the large number of pulp mills in the region as well as the proximity to Asian markets. In other areas of the U.S., however, demand for used newsprint has seen wide fluctuation.

In some U.S. states, a program called RecycleBank pays people to recycle, receiving money from local municipalities for the reduction in landfill space which must be purchased. It uses a single stream process in which all material is automatically sorted.

## Criticisms and Responses

Much of the difficulty inherent in recycling comes from the fact that most products are not designed with recycling in mind. The concept of sustainable design aims to solve this problem, and was laid out in the book *Cradle to Cradle: Remaking the Way We Make Things* by architect William McDonough and chemist Michael Braungart. They suggest that every product (and all packaging they require) should have a complete "closed-loop" cycle mapped out for each component—a way in which every component will either return to the natural ecosystem through biodegradation or be recycled indefinitely.

Complete recycling is impossible from a practical standpoint. In summary, substitution and recycling strategies only delay the depletion of non-renewable stocks and therefore may buy time in the transition to true or strong sustainability, which ultimately is only guaranteed in an economy based on renewable resources.

*—M. H. Huesemann, 2003*

While recycling diverts waste from entering directly into landfill sites, current recycling misses the dissipative components. Complete recycling is impracticable as highly dispersed wastes become so diluted that the energy needed for their recovery becomes increasingly excessive. "For example, how will it ever be possible to recycle the numerous chlorinated organic hydrocarbons that have bioaccumulated in animal and human tissues across the globe, the copper dispersed in fungicides, the lead in widely applied paints, or the zinc oxides present in the finely dispersed rubber powder that is abraded from automobile tires?"

As with environmental economics, care must be taken to ensure a complete view of the costs and benefits involved. For example, paperboard packaging for food products is more easily recycled than most plastic, but is heavier to ship and may result in more waste from spoilage.

## Energy and Material Flows

Bales of crushed steel ready for transport to the smelter

The amount of energy saved through recycling depends upon the material being recycled and the type of energy accounting that is used. Correct accounting for this saved energy can be accomplished with life-cycle analysis using real energy values. In addition, exergy, which is a measure of useful energy can be used. In general, it takes far less energy to produce a unit mass of recycled materials than it does to make the same mass of virgin materials.

Some scholars use emergy (spelled with an m) analysis, for example, budgets for the amount of energy of one kind (exergy) that is required to make or transform things into another kind of product or service. Emergy calculations take into account economics which can alter pure physics based results. Using emergy life-cycle analysis researchers have concluded that materials with large refining costs have the greatest potential for high recycle benefits. Moreover, the highest emergy efficiency accrues from systems geared toward material recycling, where materials are engineered to recycle back into their original form and purpose, followed by adaptive reuse systems where the materials are recycled into a different kind of product, and then by-product reuse systems where parts of the products are used to make an entirely different product.

The Energy Information Administration (EIA) states on its website that "a paper mill uses 40 percent less energy to make paper from recycled paper than it does to make paper from fresh lumber." Some critics argue that it takes more energy to produce recycled products than it does to dispose of them in traditional landfill methods, since the curbside collection of recyclables often requires a second waste truck. However, recycling proponents point out that a second timber or logging truck is eliminated when paper is collected for recycling, so the net energy consumption is the same. An Emergy life-cycle analysis on recycling revealed that fly ash, aluminum, recycled concrete aggregate, recycled plastic, and steel yield higher efficiency ratios, whereas the recycling of lumber generates the lowest recycle benefit ratio. Hence, the specific nature of the recycling process, the methods used to analyse the process, and the products involved affect the energy savings budgets.

It is difficult to determine the amount of energy consumed or produced in waste disposal processes in broader ecological terms, where causal relations dissipate into complex networks of material and energy flow. For example, "cities do not follow all the strategies of ecosystem development. Biogeochemical paths become fairly straight relative to wild ecosystems, with very reduced recycling, resulting in large flows of waste and low total energy efficiencies. By contrast, in wild ecosystems, one population's wastes are another population's resources, and succession results in efficient exploitation of available resources. However, even modernized cities may still be in the earliest stages of a succession that may take centuries or millennia to complete." How much energy is used in recycling also depends on the type of material being recycled and the process used to do so. Aluminium is generally agreed to use far less energy when recycled rather than being produced from scratch. The EPA states that "recycling aluminum cans, for example, saves 95 percent of the energy required to make the same amount of aluminum from its virgin source, bauxite." In 2009, more than half of all aluminium cans produced came from recycled aluminium.

Every year, millions of tons of materials are being exploited from the earth's crust, and processed into consumer and capital goods. After decades to centuries, most of these materials are "lost". With the exception of some pieces of art or religious relics, they are no longer engaged in the consumption process. Where are they? Recycling is only an intermediate solution for such materials, although it does prolong the residence time in the anthroposphere. For thermodynamic reasons, however, recycling cannot prevent the final need for an ultimate sink.

*—P. H. Brunner*

Economist Steven Landsburg has suggested that the sole benefit of reducing landfill space is trumped by the energy needed and resulting pollution from the recycling process. Others, however, have calculated through life-cycle assessment that producing recycled paper uses less energy and water than harvesting, pulping, processing, and transporting virgin trees. When less recycled paper is used, additional energy is needed to create and maintain farmed forests until these forests are as self-sustainable as virgin forests.

Other studies have shown that recycling in itself is inefficient to perform the "decoupling" of economic development from the depletion of non-renewable raw materials that is necessary for sustainable development. The international transportation or recycle material flows through "... different trade networks of the three countries result in different flows, decay rates, and potential recycling returns." As global consumption of a natural resources grows, its depletion is inevitable. The best recycling can do is to delay, complete closure of material loops to achieve 100 percent recycling of nonrenewables is impossible as micro-trace materials dissipate into the environment causing severe damage to the planet's ecosystems. Historically, this was identified as the metabolic rift by Karl Marx, who identified the unequal exchange rate between energy and nutrients flowing from rural areas to feed urban cities that create effluent wastes degrading the planet's ecological capital, such as loss in soil nutrient production. Energy conservation also leads to what is known as

Jevon's paradox, where improvements in energy efficiency lowers the cost of production and leads to a rebound effect where rates of consumption and economic growth increases.

A shop in New York only sells items recycled from demolished buildings

## Costs

The amount of money actually saved through recycling depends on the efficiency of the recycling program used to do it. The Institute for Local Self-Reliance argues that the cost of recycling depends on various factors, such as landfill fees and the amount of disposal that the community recycles. It states that communities begin to save money when they treat recycling as a replacement for their traditional waste system rather than an add-on to it and by "redesigning their collection schedules and/or trucks."

In some cases, the cost of recyclable materials also exceeds the cost of raw materials. Virgin plastic resin costs 40 percent less than recycled resin. Additionally, a United States Environmental Protection Agency (EPA) study that tracked the price of clear glass from July 15 to August 2, 1991, found that the average cost per ton ranged from $40 to $60 while a USGS report shows that the cost per ton of raw silica sand from years 1993 to 1997 fell between $17.33 and $18.10.

Comparing the market cost of recyclable material with the cost of new raw materials ignores economic externalities—the costs that are currently not counted by the market. Creating a new piece of plastic, for instance, may cause more pollution and be less sustainable than recycling a similar piece of plastic, but these factors will not be counted in market cost. A life cycle assessment can be used to determine the levels of externalities and decide whether the recycling may be worthwhile despite unfavorable market costs. Alternatively, legal means (such as a carbon tax) can be used to bring externalities into the market, so that the market cost of the material becomes close to the true cost.

## Working Conditions

The recycling of waste electrical and electronic equipment in India and China generates a significant amount of pollution. Informal recycling in an underground economy of these countries has generated an environmental and health disaster. High levels of lead (Pb), polybrominated diphenylethers (PBDEs), polychlorinated dioxins and furans, as well as polybrominated dioxins and furans (PCDD/Fs and PBDD/Fs) concentrated in the air, bottom ash, dust, soil, water, and sediments in areas surrounding recycling sites.

People in Brazil who earn their living by collecting and sorting garbage and selling them for recycling

## Environmental Impact

Economist Steven Landsburg, author of a paper entitled "Why I Am Not an Environmentalist," claimed that paper recycling actually reduces tree populations. He argues that because paper companies have incentives to replenish their forests, large demands for paper lead to large forests while reduced demand for paper leads to fewer "farmed" forests.

When foresting companies cut down trees, more are planted in their place. Most paper comes from pulp forests grown specifically for paper production. Many environmentalists point out, however, that "farmed" forests are inferior to virgin forests in several ways. Farmed forests are not able to fix the soil as quickly as virgin forests, causing widespread soil erosion and often requiring large amounts of fertilizer to maintain while containing little tree and wild-life biodiversity compared to virgin forests. Also, the new trees planted are not as big as the trees that were cut down, and the argument that there will be "more trees" is not compelling to forestry advocates when they are counting saplings.

In particular, wood from tropical rainforests is rarely harvested for paper because of their heterogeneity. According to the United Nations Framework Convention on Climate Change secretariat, the overwhelming direct cause of deforestation is subsistence farming (48% of deforestation) and commercial agriculture (32%), which is linked to food, not paper production.

## Possible Income Loss and Social Costs

In some countries, recycling is performed by the entrepreneurial poor such as the karung guni, zabbaleen, the rag-and-bone man, waste picker, and junk man. With the creation of large recycling organizations that may be profitable, either by law or economies of scale, the poor are more likely to be driven out of the recycling and the remanufacturing market. To compensate for this loss of income, a society may need to create additional forms of societal programs to help support the poor. Like the parable of the broken window, there is a net loss to the poor and possibly the whole of a society to make recycling artificially profitable e.g. through the law. However, in Brazil and Argentina, waste pickers/informal recyclers work alongside the authorities, in fully or semi-funded cooperatives, allowing informal recycling to be legitimized as a paid public sector job.

Because the social support of a country is likely to be less than the loss of income to the poor undertaking recycling, there is a greater chance the poor will come in conflict with the large recycling organizations. This means fewer people can decide if certain waste is more economically reusable in its current form rather than being reprocessed. Contrasted to the recycling poor, the efficiency of their recycling may actually be higher for some materials because individuals have greater control over what is considered "waste."

One labor-intensive underused waste is electronic and computer waste. Because this waste may still be functional and wanted mostly by those on lower incomes, who may sell or use it at a greater efficiency than large recyclers.

Some recycling advocates believe that laissez-faire individual-based recycling does not cover all of society's recycling needs. Thus, it does not negate the need for an organized recycling program. Local government can consider the activities of the recycling poor as contributing to property blight.

## Public Participation Rates

Changes that have been demonstrated to increase recycling rates include:

- Single-stream recycling
- Pay as you throw fees for trash

"Between 1960 and 2000, the world production of plastic resins increased 25-fold, while recovery of the material remained below 5 percent." Many studies have addressed recycling behaviour and strategies to encourage community involvement in recycling programmes. It has been argued that recycling behaviour is not natural because it requires a focus and appreciation for long-term planning, whereas humans have evolved to be sensitive to short-term survival goals; and that to overcome this innate predisposition, the best solution would be to use social pressure to compel participation in recycling programmes. However, recent studies have concluded that social pressure is unviable in this context. One reason for this is that social pressure functions well in small group sizes of 50 to 150 individuals (common to nomadic hunter–gatherer peoples) but not in communities numbering in the millions, as we see today. Another reason is that individual recycling does not take place in the public view.

In a study done by social psychologist Shawn Burn, it was found that personal contact with individuals within a neighborhood is the most effective way to increase recycling within a community. In his study, he had 10 block leaders talk to their neighbors and persuade them to recycle. A comparison group was sent fliers promoting recycling. It was found that the neighbors that were personally contacted by their block leaders recycled much more than the group without personal contact. As a result of this study, Shawn Burn believes that personal contact within a small group of people is an important factor in encouraging recycling. Another study done by Stuart Oskamp examines the effect of neighbors and friends on recycling. It was found in his studies that people who had friends and neighbors that recycled were much more likely to also recycle than those who didn't have friends and neighbors that recycled.

Many schools have created recycling awareness clubs in order to give young students an insight on recycling. These schools believe that the clubs actually encourage students to not only recycle at school but at home as well.

## Reuse

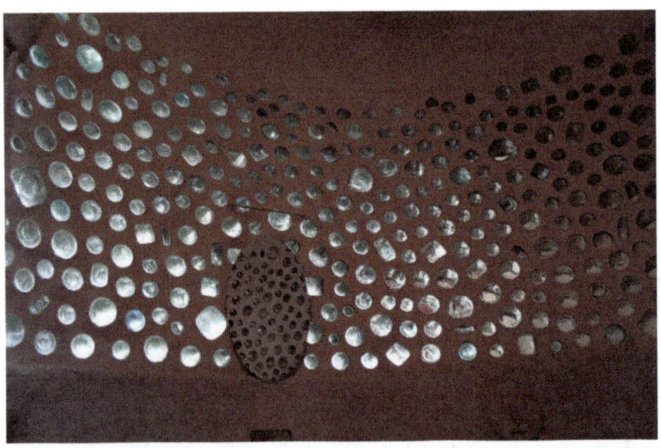

An interior bathroom wall that incorporates repurposed clear glass bottles into a bottle wall

Reuse is the action or practice of using something again, whether for its original purpose (conventional reuse) or to fulfil a different function (creative reuse or repurposing). It should be distinguished from recycling, which is the breaking down of used items to make raw materials for the manufacture of new products. Reuse – by taking, but not reprocessing, previously used items

– helps save time, money, energy, and resources. In broader economic terms, it can make quality products available to people and organizations with limited means, while generating jobs and business activity that contribute to the economy.

Historically, financial motivation was one of the main drivers of reuse. In the developing world this driver can lead to very high levels of reuse, however rising wages and consequent consumer demand for the convenience of disposable products has made the reuse of low value items such as packaging uneconomic in richer countries, leading to the demise of many reuse programs. Current environmental awareness is gradually changing attitudes and regulations, such as the new packaging regulations, are gradually beginning to reverse the situation.

One example of conventional reuse is the doorstep delivery of milk in refillable bottles; other examples include the retreading of tires and the use of returnable/reusable plastic boxes, shipping containers, instead of single-use corrugated fiberboard boxes.

## Advantages and Disadvantages

Ants, being social insects, have been able to reuse rail tracks abandoned by humans for their own transportation. (Kadina, South Australia)

Reuse has certain potential advantages:

- Energy and raw materials savings as replacing many single use products with one reusable one reduces the number that need to be manufactured.
- Reduced disposal needs and costs.
- Refurbishment can bring sophisticated, sustainable, well paid jobs to underdeveloped economies.
- Cost savings for business and consumers as a reusable product is often cheaper than the many single use products it replaces.
- Some older items were better handcrafted and appreciate in value.

Disadvantages are also apparent:

- Reuse often requires cleaning or transport, which have environmental costs.

- Some items, such as freon appliances, infant auto seats, older tube TVs and secondhand automobiles could be hazardous or less energy efficient as they continue to be used.
- Reusable products need to be more durable than single-use products, and hence require more material per item. This is particularly significant if only a small proportion of the reusable products are in fact reused.
- Sorting and preparing items for reuse takes time, which is inconvenient for consumers and costs money for businesses.
- Special skills are required to tweak the functional throughput of items when devoting them to new uses outside of their original purpose.
- Knowing the standards that legacy products conform to is required for knowing what adapters to buy for newer products to be compatible with them, even though the cost of adapters for such applications is a minor disadvantage.
- Being a rather minor disadvantage, metal that is repurposed later on can sometimes contain rust, seeing as it sometimes ages before reuse.

## Examples

### Reuse Centers and Virtual Exchange

A salvaged window from the deconstruction of an old house turned home decor with paint and stencils. Source: Habitat for Humanity Mt. Angel ReStore

An electric wire reel reused as a center table in a Rio de Janeiro decoration fair. The reuse of materials is a very sustainable practice that is rapidly growing among designers in Brazil

These services facilitate the transaction and redistribution of unwanted, yet perfectly usable, materials and equipment from one entity to another. The entities that benefit from either side of this service (as donors, sellers, recipients, or buyers) can be businesses, nonprofits, schools, community groups, and individuals. Some maintain a physical space (a reuse center), and others act as a matching service (a virtual exchange). Reuse centers generally maintain both warehouses and trucks. They take possession of the donated materials and make them available for redistribution or sale. Virtual exchanges do not have physical space or trucks, but instead allow users to post listings of materials available and wanted (for free or at low cost) on an online materials exchange website. Staff will help facilitate the exchange of these materials without ever taking possession of the materials.

United States

- Goodwill Industries
- Salvation Army
- Second Harvest Food Bank
- Habitat for Humanity ReStores. Virtual exchangees include:
    - CalMax (California)
    - WasteMatch (New York)
- Alchemy Goods
- Materials for the Arts (Queens, New York)
- STAY VOCAL (Norwell, MA)

Consumer resources exist for exchanging usable materials, such as freecycling sites which are often grassroots and entirely nonprofit movements of people who are giving (and getting) stuff for free in their own towns. It's all about reuse and keeping good stuff out of landfills. Membership is normally free. In, addition, there are directory-based resources such as RecycleChicken.com which point consumers to local and national locations for reuse and repurposing of materials not normally accepted in recycling programs.

- Teleplan Camera Repair has introduced a free camera recycling program through the reuse of cameras.

Australia

- Woolloongabba, Queensland
    - Reverse Garbage Queensland
- Marrickville, New South Wales:
    - Reverse Garbage is the largest reuse centre in the country, diverting more than 12,000 cubic metres of resources from landfills each year.

  - The Bower Reuse and Repair Centre diverts more than 7,500 cubic metres of 'waste' from landfills a year in a building entirely made of salvaged materials.
- Perth, Western Australia:
  - REmida WA

With technological innovations, new applications and shorter product lifetimes there is an ever increasing quantity of waste electrical and electronic equipment. The environmental pressures associated with this are well documented and include material and energy losses and an increase in air, water and land pollution from waste treatment methods. The incorrect disposal and treatment of electrical and electronic equipment also poses threats to human health, particularly when involving illegal exports.

## Addressing Issues of Repair, Reuse and Recycling

One way to address this is to increase product longevity; either by extending a product's first life or addressing issues of repair, reuse and recycling. Reusing products, and therefore extending the use of that item beyond the point where it is discarded by its first user is preferable to recycling or disposal, as this is the least energy intensive solution, although it is often overlooked.

The EU Circular Economy Package recognises the importance of extending product lifetimes and includes repair and reuse of products in its action plan to ensure products reach their optimum lifespan. If targets for reducing greenhouse gas emissions are to be reached, then reuse needs to be included as part of a whole life cycle approach.

A strong second hand market-place exists, with charity shops on most high streets, car boot sales and online auction sites maintaining popularity and regular TV shows featuring both buying and selling at auction.

## Inadequate Repair Infrastructure

However, the reuse of electrical and electronic equipment products remains low due to inadequate infrastructure, lack of repair networks and products. Local authority collection systems are better suited for handling waste than handling goods and preserving reuse potential. Retailer delivery staff are trained to handle goods carefully.

So, do we need a radical rethink on how we move unwanted, still useable items to the second-hand market place? Is there a case for changing our approach to producer responsibility and insisting that producers finance collection for reuse, and additionally, drive consumer choices for reuse, repair and remanufacture; whilst addressing the costs of recycling and disposal?

There are opportunities for producers, waste management companies and local authorities to make both repair and reuse habitual, whilst these require changes to householder behaviour change through raising awareness, they also require investment in infrastructure and logistical operations. Is it time to insist that more products are designed to have longer lifetimes? That they can be disassembled, repaired and reused before being recycled?

## Business Models Providing Opportunities

This would not necessarily be a poor strategy for businesses, there are business models that provide opportunities to retain ownership of valuable products and components through leasing, servicing, repair and re-sale.

While it is choices made by consumers that will ultimately determine the success of such ventures, there is huge potential for the reuse of goods and materials to deliver social and economic and environmental benefits. The EU Circular Economy Package, the Scottish Circular Economy Strategy and the national reuse target set by the Spanish Government are examples of governments recognising that second-hand goods should be a good value mainstream option and are working towards making reuse easier for consumers.

In environmental terms, reuse ought to be more common than recycling and energy recovery, with both the financial and environmental costs of simple refurbishment of some products being a fraction of original manufacturing costs. If we are going to be serious about living in a Circular Economy we need to recognise the value of our waste and ensure resources are kept in the economy for longer, slow down the use of valuable raw materials and ensure that products are reused and materials are recycled rather than landfilled.

## Remanufacturing

The most involved reuse organizations are "repair and overhaul" industries which take valuable parts, such as engine blocks, office furniture, toner cartridges, single-use cameras, aircraft hulls, and cathode ray tubes (CRTs) and refurbish them in a factory environment in order to meet the same/similar specifications as new products. Xerox (copy machines), and Cummins Engine are examples of refurbishing factories in the USA. Rolls Royce has a very large aircraft remanufacturing factory in Singapore; Caterpillar recently announced the opening of a tractor refurbishing plant in China. Some factories operate in competition with the original equipment manufacturer (OEM). When the refurbished item is resold under a new label (used monitor CRTs made into TVs, or cameras resold under a new label) this has been found legal by most courts.

When the item is resold under the same OEM name, it is informally considered a "gray market" item - if it is sold as used, it's legal, if it's represented as an OEM product eligible for rebates and warranties, it is considered "counterfeit" or "black market". The automobile parts industry in the USA is governed by laws on the disclosure of "used" parts and, in some states, mattresses which have been used are required to be sanitized or destroyed. Whether these laws are in place to protect consumers from black market items, or to protect manufacturers ("hindsight obsolescence"), is often an area of intense debate. Fuji Photo Film Co. v. Jazz Photo Corp. is a recent example of the war between patent holders and refurbishing factories. To quote the 2003 District Court of New Jersey:*"Thus, the key issue in the dispute between Fuji and Jazz is whether the cameras sold by Jazz are "refurbished" in such a way that they can be considered to have been permissibly "repaired" or impermissibly "reconstructed."*

## Package Deposit Programs

Deposit programs offer customers a financial incentive to return packaging for reuse. Although no longer common, international experience is showing that they can still be an effective way to encourage packaging reuse. However, financial incentive, unless great, may be less of an incentive than convenience: statistics show that, on average, a milk bottle is returned 12 times, whereas a lemonade bottle with a 15p deposit is returned, on average, only 3 times.

Reusable glass bottles collected in Bishkek, Kyrgyzstan. Deposit values (0.5-2 Kyrgyz som) are posted next to the sample bottles on the rack

Refillable bottles are used extensively in many European countries; for example in Denmark, 98% of bottles are refillable, and 98% of those are returned by consumers. These systems are typically supported by deposit laws and other regulations.

Sainsbury Ltd have operated a plastic carrier bag cash refund scheme in 1991 - "the penny back scheme". The scheme is reported to save 970 tonnes of plastic per annum. The scheme has now been extended to a penny back on a voucher which can be contributed to schools registered on the scheme; it estimates this will raise the savings in plastic to 2500 tonnes per annum.

The 600 ml brown bottle is the "standard beer reused bottle" in Brazil.

In some developing nations like India and Pakistan, the cost of new bottles often forces manufacturers to collect and refill old glass bottles for selling cola and other drinks. India and Pakistan also have a way of reusing old newspapers: "Kabadiwalas" buy these from the readers for scrap value and reuse them as packaging or recycle them. Scrap intermediaries help consumer dispose of other materials including metals and plastics.

## Closed-loop Programs

These apply primarily to items of packaging, for example, where a company is involved in the regular transportation of goods from a central manufacturing facility to warehouses or warehouses to retail outlets. In these cases there is considerable benefit to using reusable "transport packaging" such as plastic crates or pallets.

The benefits of closed-loop reuse are primarily due to low additional transport costs being involved, the empty lorry returning with the empty crates. There have been some recent attempts to get the public to join in on closed loop reuse schemes where shoppers use reusable plastic baskets in place of carrier bags for transporting their goods home from the supermarket; these baskets fit on specially designed trolleys making shopping supposedly easier.

## Refilling Programs

There have been some market-led initiatives to encourage packaging reuse by companies introducing refill packs of certain commodities (mainly soap powders and cleaning fluids), the contents being transferred before use into a reusable package kept by the customer, with the savings in packaging being passed onto the customer by lower shelf prices. The refill pack itself is not reused, but being a minimal package for carrying the product home, it requires less material than one with the durability and features (reclosable top, convenient shape, etc.) required for easy use of the product, while avoiding the transport cost and emissions of returning the reusable package to the factory.

## Regifting

Some items, such as clothes and children's toys, often become unwanted before they wear out due to changes in their owner's needs or preferences; these can be reused by selling or giving them to new owners. Regiving can take place informally between family, friends, or neighbours, through environmental freecycling organisations or through anti-poverty charities such as the Red Cross, United Way, Salvation Army, and Goodwill which give these items to those who could not afford them new. Other organizations such as iLoveSchools have websites where both new and used goods can be offered to any of America's school teachers so their life can be extended and help schoolchildren. The average American, for example, throws away 67.9 pounds of used clothing and rags. With the U.S. population at approximately 296 million people, that translates into 20 billion pounds of used clothing and textiles that are tossed into the landfills each year. This has partly motivated movements such as The Compact, whose members promise not to buy anything new for a year, and rely on reusing items that otherwise would be thrown away. Reuse not only reduces landfill inline with the waste minimization program but can help raise money for a good cause.

## Printer Cartridges and Toners

Printer ink cartridges can be reused. They are sorted by brand and model, to be refilled or resold back to the manufacturers. The companies then refill the ink reservoir to resell to consumers. Toner cartridges are recycled the same way as ink cartridges, using toner instead of ink. This method is highly efficient as there is no energy spent on melting and recreating the cartridges.

## Repurposing

Repurposing is to use a tool for use as another tool, usually for a purpose unintended by the original tool-maker. Typically, repurposing is done using items usually considered to be junk or garbage. A good example of this would be the Earthship style of house, that uses tires as insulating walls and bottles as glass walls. Reuse is not limited to repeated uses for the same purpose. Examples of re-purposing include using tires as boat fenders and steel drums or plastic drums as feeding troughs and/or composting bins.

## Waste Exchanges

A waste exchange, or virtual exchange, facilitates the use of a waste product from one process as a raw material for another. As with new life reuse of finished items, this avoids the environmental costs of disposing of the waste and obtaining new raw material, and may still be possible if the nature of the process makes avoiding production of the waste or recycling it back into the original process impossible.

This sort of scheme needs to have a far broader base than is currently the case, it requires organization and the setting up of waste brokerages where lists of currently available wastes are and the quantities available. One of the problems is once a demand for a waste is known or shown then the material is no longer a "waste" but a sellable commodity which often prices itself out of the market, c.f waste cement kiln dust and N-viro (lime conditioned sewage sludge fertilizer). In the former East Germany, organic household waste was collected and used as fodder for pigs. This integrated system was made possible by the state's control of agriculture; the complexities of continuing it in a market economy after German reunification meant the system had to be discontinued.

## Reuse of Waste Water and Excreta in Agriculture

The nutrients, i.e. nitrogen, phosphorus, potassium and micronutrients, and organic matter contained in wastewater, excreta (urine and feces) and greywater have traditionally been reused in agriculture in many countries and are still being reused in agriculture to this day - unfortunately often in an unregulated and unsafe manner for example in many developing countries (e.g. Mexico, India, Bangladesh, Ghana). The WHO Guidelines from 2006 have set up a framework how this reuse can be done safely by following a multiple barrier approach. Work by the International Water Management Institute has led to a better understanding on how such wastewater reuse can be safely implemented in practice, for which they won the Stockholm Water Prize in 2012. Reuse of sanitised excreta in agriculture has also been called a "closing the loop" approach for sanitation and agriculture and is central to the ecological sanitation approach.

## Reuse of Musical Instruments

Reuse of pianos for street food promotion

## Measuring the Impact of Reuse, Reuse Metrics

Determining the balance of how the several effects of reuse interact is often best accomplished by a formal life cycle assessment. For example, research has shown that reusing a product can reduce $CO_2$ emissions and carbon footprint by more than 50% relative to the complete product life cycle. A relatively unknown effective way to reduce $CO_2$ emissions and carbon footprint is reusing products. Often the relative carbon footprint of manufacturing and the supply chain is unknown. A scientific methodology has been developed to calculate how much $CO_2$ emissions are reduced when buying used or second hand hardware versus new hardware, the so-called durability greener network calculator.

There are many ways of measuring the positive environmental, economic and social impact data. These include:

- # of tons diverted from the landfill
- $ avoided disposal costs
- $ avoided purchase costs
- $ value of materials donated
- $ revenues earned
- # of jobs created or retained
- # of families/individuals/organizations assisted

## Internalized Environmental Costs

A Pigovian tax is an environmental tax: a charge on items that reflects the environmental costs of their manufacture and disposal. This makes the environmental benefit of using one reusable item instead of many disposable ones into a financial incentive. Such charges have been introduced in some countries.

## Comparison to Recycling

A school being prepared for reuse as housing

Recycling differs from reuse in that it breaks down the item into raw materials which are then used to make new items, as opposed to reusing the intact item. As this extra processing requires energy, as a rule of thumb reuse is environmentally preferable to recycling ("reduce, reuse, recycle"), though recycling does have a significant part to play as it can often make use of items which are broken, worn out or otherwise unsuitable for reuse. However, as transport emissions are significant portion of the environmental impact of both reuse and recycling, in some cases recycling is the more prudent course as reuse can require long transport distances. A complex life cycle analysis may be required during a product's design phase to determine the efficacy of reuse, recycling, or neither, and produce accordingly.

### Reuse of Information

Besides physical resources, information is often reused, notably program code for thesoftware that drives computers and the Internet, but also the documentation that explains how to use every modern device. And it is proposed as a way to improve education by assembling a great library of shareable learning objects that can be reused in learning management systems.

Software reuse grew out of the standard subroutine libraries of the 1960s. It is the main principle of today's object-oriented programming. Instead of constantly reinventing software wheels, programming languages like C++, Java, Objective-C, and others are building vast collections of reusable software objects and components.

Reuse of information has a tremendous return on investment for organizations whose documentation is translated into many languages. Translation memory systems can store text that has already been translated into dozens of languages for retrieval and reuse.

### Reuse of Older Software

Reuse of older software is popular among retrocomputing. Most of the time, emulators are used to run older software from other platforms, or other operating systems.

Sometimes, older operating systems such as DOS are reused for computing roles that don't demand lots of computing power. However, the widespread of availability of secondhand Windows XP computers at extreme low prices has largely supplanted immediate opportunities for using DOS on some repurposing applications, especially since something like USB isn't found on most pre-Windows XP computers.

## Waste Minimisation

Waste minimisation is a process of elimination that involves reducing the amount of waste produced in society and helps to eliminate the generation of harmful and persistent wastes, supporting the efforts to promote a more sustainable society. Waste minimisation involves redesigning products and/or changing societal patterns, concerning consumption and production, of waste generation, to prevent the creation of waste.

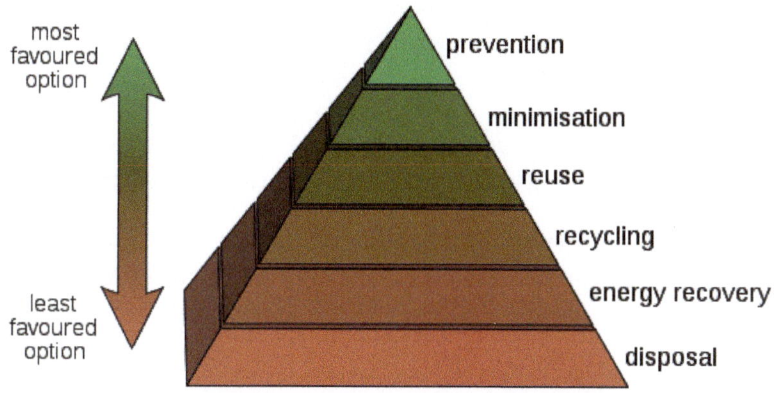


Waste hierarchy. Refusing, reducing, reusing, recycling and composting allow to reduce waste.

The most environmentally resourceful, economically efficient, and cost effective way to manage waste is to not have to address the problem in the first place. Waste minimisation should be seen as a primary focus for most waste management strategies. Proper waste management can require a significant amount of time and resources; therefore, it is important to understand the benefits of waste minimisation and how it can be implemented in all sectors of the economy, in an effective, safe and sustainable manner.

The basic concept behind waste management is the waste hierarchy, where the most effective approaches to managing waste are at the top. Waste management is in contrast to waste minimisation. Waste management focuses on processing waste after it is created, concentrating on re-use, recycling, and waste-to-energy conversion rather than eliminating the creation of waste in the initial phases of production. Waste minimisation involves efforts to minimize resource and energy use during manufacture. For the same commercial output, usually the less materials are used, the less waste is produced. Waste minimisation usually requires knowledge of the production process, cradle-to-grave analysis (the tracking of materials from their extraction to their return to earth) and detailed knowledge of the composition of the waste.

The main sources of waste vary from country to country. In the UK, most waste comes from the construction and demolition of buildings, followed by mining and quarrying, industry and commerce. Household waste constitutes a relatively small proportion of all waste. Reasons for the creation of waste sometimes include requirements in the supply chain. For example, a company handling a product may insist that it should be packaged using particular packing because it fits its packaging equipment.

## Benefits

Waste minimisation can protect the environment and provide good economic and business practices. Waste minimisation can improve:

- Efficient production practices. Waste minimisation can achieve more output of product per unit of input of raw materials.
- Economic returns. More efficient use of products means reduced costs of purchasing new materials improving the financial performance of a company.

- Public image. The environmental profile of a company is an important part of its overall reputation and waste minimisation reflects a proactive movement towards environmental protection.
- Quality of products produced. New innovation and technological practices can reduce waste generation and improve the quality of the inputs in the production phase.
- Environmental responsibility. Minimising or eliminating waste generation makes it easier to meet targets of environmental regulations, policies, and standards. The environmental impact of waste will be reduced.

## Industries

In industries, using more efficient manufacturing processes and better materials will generally reduce the production of waste. The application of waste minimisation techniques has led to the development of innovative and commercially successful replacement products. Waste minimisation has proven benefits to industry and the wider environment. It helps in value creation and increase quality of work.

Waste minimisation often requires investment, which is usually compensated by the savings. However, waste reduction in one part of the production process may create waste production in another part.

There are government incentives for waste minimisation, which focus on the environmental benefits of adopting waste minimisation strategies. In the UK, several pilot schemes such as The Catalyst Project and the Dee Waste Minimisation Project, have shown the efficacy of such policies. Fourteen companies in Merseyside took part in the Catalyst Project; the project generated overall savings of £9 million and landfill waste was reduced by 12,000 tonnes per year.

## Processes

- Resource optimization

  Minimising the amount of waste produced by organisations or individuals goes hand-in-hand with optimizing their use of raw materials. For example, a dressmaker may arrange pattern pieces on a length of fabric in a particular way to enable the garment to be cut out from the smallest area of fabric.

- Reuse of scrap material

  Scraps can be immediately re-incorporated at the beginning of the manufacturing line so that they do not become a waste product. Many industries routinely do this; for example, paper mills return any damaged rolls to the beginning of the production line, and in the manufacture of plastic items, off-cuts and scrap are re-incorporated into new products.

- Improved quality control and process monitoring

  Steps can be taken to ensure that the number of reject batches is kept to a minimum. This is achieved by increasing the frequency of inspection and the number of points of inspection.

For example, installing automated continuous monitoring equipment can help to identify production problems at an early stage.

- Waste exchanges

  This is where the waste product of one process becomes the raw material for a second process. Waste exchanges represent another way of reducing waste disposal volumes for waste that cannot be eliminated.

- Ship to point of use

  This involves making deliveries of incoming raw materials or components direct to the point where they are assembled or used in the manufacturing process to minimise handling and the use of protective wrappings or enclosures.

- Zero waste

  This is a whole systems approach that aims to eliminate waste at the source and at all points down the supply chain, with the intention of producing no waste. It is a design philosophy which emphasizes waste prevention as opposed to end of pipe waste management.

## Product Design

Waste minimisation and resource maximisation for manufactured products can most easily be done at the design stage. Reducing the number of components used in a product or making the product easier to take apart can make it easier to be repaired or recycled at the end of its useful life.

In some cases, it may be best not to minimise the volume of raw materials used to make a product, but instead reduce the volume or toxicity of the waste created at the end of a product's life, or the environmental impact of the product's use.

## Fitting the Intended Use

In this strategy, products and packages are optimally designed to meet their intended use. This applies especially to packaging materials, which should only be as durable as necessary to serve their intended purpose. On the other hand, it could be more wasteful if food, which has consumed resources and energy in its production, is damaged and spoiled because of extreme measures to reduce the use of paper, metals, glass and plastics in its packaging.

## Durability

Improving product durability, such as extending a vacuum cleaner's useful life to 15 years instead of 12, can reduce waste and usually much improves resource optimisation.

But in some cases it has a negative environmental impact. If a product is too durable, its replacement with more efficient technology is likely to be delayed. For example, a washing machine produced 10 years ago may use twice as much water, detergent and energy as one produced today. Therefore, extending an older machine's useful life may place a heavier burden on the environment than scrapping it, recycling its metal and buying a new model. Similarly, older vehicles consume more fuel and produce more emissions than their modern counterparts.

Most proponents of waste minimisation consider that the way forward may be to view any manufactured product at the end of its useful life as a resource for recycling and reuse rather than waste. Recycling a product is easier if it is constructed of fewer materials. Car manufacturers have recently reduced the number of plastics used in their cars from twenty or more to three or four, hence simplifying the recovery of plastics from scrapped cars. However, exceptions (like having a combination of paper and plastic or plastic coating on glass) do exist, and might enable a product to fulfill its role with the minimum of resources.

Making refillable glass bottles strong enough to withstand several journeys between the consumer and the bottling plant requires making them thicker and so heavier, which increases the resources required to transport them. Since transport has a large environmental impact, careful evaluation is required of the number of return journeys bottles make. If a refillable bottle is thrown away after being refilled only several times, the resources wasted may be greater than if the bottle had been designed for a single journey..

Many choices involve trade-offs of environmental impact, and often there is insufficient information to make informed decisions.

## Households

Appropriate amounts and sizes can be chosen when purchasing goods; buying large containers of paint for a small decorating job or buying larger amounts of food than can be consumed create unnecessary waste. Also, if a pack or can is to be thrown away, any remaining contents must be removed before the container can be recycled.

Home composting, the practice of turning kitchen and garden waste into compost can be considered waste minimisation.

The resources that households use can be reduced considerably by using electricity thoughtfully (e.g. turning off lights and equipment when it is not needed) and by reducing the number of car journeys made. Individuals can reduce the amount of waste they create by buying fewer products and by buying products which last longer. Mending broken or worn items of clothing or equipment also contributes to minimising household waste. Individuals can minimise their water usage, and walk or cycle to their destination rather than using their car to save fuel and cut down emissions.

In a domestic situation, the potential for minimisation is often dictated by lifestyle. Some people may view it as wasteful to purchase new products solely to follow fashion trends when the older products are still usable. Adults working full-time have little free time, and so may have to purchase more convenient foods that require little preparation, or prefer disposable nappies if there is a baby in the family.

The amount of waste an individual produces is a small portion of all waste produced by society, and personal waste reduction can only make a small impact on overall waste volumes. Yet, influence on policy can be exerted in other areas. Increased consumer awareness of the impact and power of certain purchasing decisions allows industry and individuals to change the total resource consumption. Consumers can influence manufacturers and distributors by avoiding buying products that do not have eco-labelling, which is currently not mandatory, or choosing products that minimise the use of packaging. In the UK, PullApart combines both environmental and consumer packaging surveys, in a curbside packaging recycling classification system to minimise waste. Where reuse schemes are available, consumers can be proactive and use them.

## Health-care Facilities

Health-care establishments are massive producers of waste. The major sources of health-care waste are: hospitals, laboratories and research centres, mortuary and autopsy centres, animal research and testing laboratories, blood banks and collection services, and nursing homes for the elderly.

Waste minimisation can offer many opportunities to these establishments to use fewer resources, be less wasteful and generate less hazardous waste. Good management and control practices among health-care facilities can have a significant effect on the reduction of waste generated each day.

## Practices

There are many examples of more efficient practices that can encourage waste minimization in healthcare establishments and research facilities

## Source Reduction

- Purchasing reductions which ensures the selection of supplies that are less wasteful or less hazardous.
- The use of physical rather than chemical cleaning methods such as steam disinfection instead of chemical disinfection.
- Preventing the unnecessary wastage of products in nursing and cleaning activities.

## Management and Control Measures at Hospital Level

- Centralized purchasing of hazardous chemicals.
- Monitoring the flow of chemicals within the health care facility from receipt as a raw material to disposal as a hazardous waste.
- The careful separation of waste matter to help minimize the quantities of hazardous waste and disposal.

## Stock management of Chemical and Pharmaceutical Products

- Frequent ordering of relatively small quantities rather than large quantities at one time.
- Using the oldest batch of a product first to avoid expiration dates and unnecessary waste.
- Using all the contents of a container containing hazardous waste.
- Checking the expiry date of all products at the time of delivery.

# Environmental Mitigation

Environmental mitigation, compensatory mitigation, or mitigation banking, are terms used primarily by the United States government and the related environmental industry to describe projects or programs intended to offset known impacts to an existing historic or natural resource such as a stream, wetland, endangered species, archeological site or historic structure. To "mitigate" means to make less harsh or hostile. Environmental mitigation is typically a part of an environmental crediting system established by governing bodies which involves allocating debits and credits. Debits occur in situations where a natural resource has been destroyed or severely impaired and credits are given in situations where a natural resource has been deemed to be improved or preserved. Therefore, when an entity such as a business or individual has a "debit" they are required to purchase a "credit". In some cases credits are bought from "mitigation banks" which are large mitigation projects established to provide credit to multiple parties in advance of development when such compensation cannot be achieved at the development site or is not seen as beneficial to the environment. Crediting systems can allow credit to be generated in different ways. For example, in the United States, projects are valued based on what the intentions of the project are which may be to preserve, enhance, restore or create (PERC) a natural resource.

## Advantages

Environmental mitigation and crediting systems are often praised for the following reasons:

### Development-friendly

Mitigation is a more development-friendly alternative to strict environmental laws because it allows development to occur where environmental laws might prohibit it.

### Mitigation Industry

Mitigation inevitably creates a "mitigation industry". By requiring those who impact natural resources to purchase credits, a demand for mitigation credit is formed. Businesses related to environmental work typically benefit from such a system.

### Targeting Ecological Value

Mitigation has the potential to save and restore the most valuable environmental resources at the least cost, assuming that regulation 1) protects health and welfare as defined by the National Environmental Policy Act (NEPA) and 2) assures that a credit accurately represents measurable ecological value. Buyers are typically looking for mitigation credits that are both cheap and the most likely to meet regulatory requirements for compensatory mitigation. Regulators must therefore find a balance between protecting the long term public interest and ensuring that buyers have the proper incentives to participate in the environmental marketplace.

### Cost Burden

Mitigation systems place the environmental costs of development mostly on the individuals or entities that are impacting the environment. Without environmental mitigation, costs of alleviating environmental damage caused by development could be placed in the hands of the government which would in turn pass costs on to taxpayers not responsible for environmental impacts.

## Benefit to Landowners

Land previously unused or impractical for development is given greater monetary value under a mitigation system. For instance, land in floodplains may be impractical for commercial or residential development but conductive for mitigation activities. Land in rural areas with very little potential for growth are more valuable when given the opportunity to be used for mitigation credits.

## Disadvantages

The following are criticisms of environmental mitigation and crediting systems:

## Incorrect Allocation and Valuation of Credits and Debits

Mitigation regulations may not properly take into account the total ecological losses and gains associated with environmental impacts or mitigation when allocating debits and credits. Governing bodies are primarily responsible for prescribing the ecological criteria required to attain credits for mitigation. They are also responsible for valuation of credit. Therefore, it is evident that problems with the allocation and valuation of credits and debits might stem from the complexity of assessing the current comparative value of ecological resources (aka ecosystem services), ecosystem change over time, and/or a lack of understanding about what is beneficial or harmful to the environment overall. To address these uncertainties regulators often assign 'coverage ratios' to compensatory mitigation agreements. Coverage ratios of, for example, 3:1 require 3 compensatory mitigation credits for every 1 unit of ecological disturbance.

## Effects on Land Cost and Availability

Mitigation could be seen as contributing to the increasing cost of land because some mitigation work requires that large amounts of land be purchased or put into conservation easements. Mitigation can therefore compete with other rural land uses such as agriculture and residential development. This suggests that land owners must be alert to find the highest and best use for their properties given the potential market value that mitigation credits represent.

## 'In Perpetuity' Commitments of Land

Commitment of lands to compensatory mitigation must be done 'in perpetuity', meaning permanently into the future. Otherwise, the long-term public interest could not be served via compensatory mitigation programs. This means that properties must continue to be managed with ecosystem values in mind, sometimes preventing landowners from transforming the landscape to meet changing needs. For example, future large scale development projects would not likely be permitted on previously dedicated mitigation properties.

# Regulation and Monitoring of Pollution

To protect the environment from the adverse effects of pollution, many nations worldwide have enacted legislation to regulate various types of pollution as well as to mitigate the adverse effects of pollution.

## Regulation and Monitoring by Region

### International

Since pollution crosses political boundaries international treaties have been made through the United Nations and its agencies to address international pollution issues.

### Greenhouse Gas Emissions

The Kyoto Protocol is an amendment to the United Nations Framework Convention on Climate Change (UNFCCC), an international treaty on global warming. It also reaffirms sections of the UNFCCC. Countries which ratify this protocol commit to reduce their emissions of carbon dioxide and five other greenhouse gases, or engage in emissions trading if they maintain or increase emissions of these gases. A total of 141 countries have ratified the agreement. Notable exceptions include the United States and Australia, who have signed but not ratified the agreement. The stated reason for the United States not ratifying is the exemption of large emitters of greenhouse gases who are also developing countries, like China and India.

An UN environmental conference held in Bali 3–14 December 2007 with the participation from 180 countries aims to replace the Kyoto Protocol, which will end in 2012. During the first day of the conference United States, Saudi Arabia and Canada were presented with the "Fossil-of-the-day-award", a symbolic bag of coal for their negative impact on the global climate. The bags included the flags of the respective countries.

### Toxic Wastes

### Canada

In Canada the regulation of pollution and its effects are monitored by a number of organizations depending on the nature of the pollution and its location. The three levels of government (Federal – Canada Wide; Provincial; and Municipal) equally share in the responsibilities, and in the monitoring and correction of pollution.

### China

China's rapid industrialization has substantially increased pollution. China has some relevant regulations: the 1979 Environmental Protection Law, which was largely modeled on U.S. legislation, but the environment continues to deteriorate. Twelve years after the law, only one Chinese city was making an effort to clean up its water discharges. This indicates that China is about 30 years behind the U.S. schedule of environmental regulation and 10 to 20 years behind Europe. In July 2007, it was reported that the World Bank reluctantly censored a report revealing that 750,000 people in China die every year as a result of pollution-related diseases. China's State Environment Protection Agency and the Health Ministry asked the World Bank to cut the calculations of premature deaths from the report fearing the revelation would provoke "social unrest".

## Europe

The basic European rules are included in the Directive 96/61/EC of 24 September 1996 concerning integrated pollution prevention and control (IPPC) and the National Emission Ceilings Directive.

## United Kingdom

In the 1840s, the United Kingdom brought onto the statute books legislation to control water pollution. It was extended to all rivers and coastal water by 1961. However, currently the clean up of historic contamination is controlled under a specific statutory scheme found in Part IIA of the Environmental Protection Act 1990 (Part IIA), as inserted by the Environment Act 1995, and other 'rules' found in regulations and statutory guidance. The Act came into force in England in April 2000.

Within the current regulatory framework, Pollution Prevention and Control (PPC) is a regime for controlling pollution from certain industrial activities. The regime introduces the concept of Best Available Techniques ("BAT") to environmental regulations. Operators must use the BAT to control pollution from their industrial activities to prevent, and where that is not practicable, to reduce to acceptable levels, pollution to air, land and water from industrial activities. The Best Available Techniques also aim to balance the cost to the operator against benefits to the environment. The system of Pollution Prevention and Control is replacing that of Integrated Pollution Control (IPC) (which was established by the Environmental Protection Act 1990) and is taking effect between 2000 and 2007. The Pollution Prevention and Control regime implements the European Directive (EC/96/61) on integrated pollution prevention and control.

## United States

A polluted ditch along Interstate 25 between Colorado Springs and Pueblo, Colorado.

## Air Pollution

The United States Congress passed the Clean Air Act in 1963 to legislate the reduction of smog and atmospheric pollution in general. That legislation has subsequently been amended and extended in 1966, 1970, 1977 and 1990. Numerous state and local governments have enacted similar legislation either implementing or filling in locally important gaps in the national program. The national Clean Air Act and similar state legislative acts have led to the widespread use of atmospheric dispersion modeling in order to analyze the air quality impacts of proposed major actions. With the 1990 Clean Air Act, the United States Environmental Protection Agency (EPA) began a controversial carbon trading system in which tradable rights to emit a specified level of carbon are granted to polluters.

## Water Pollution

Enactment of the 1972 Clean Water Act required thousands of facilities to obtain permits for discharges to navigable waters, through the National Pollutant Discharge Elimination System (NPDES). It also required EPA to establish national discharge standards for municipal sewage treatment plants, and for many industrial categories (the latter are called "effluent guidelines.")

Municipal and industrial permittees are required to regularly collect and analyze wastewater samples, and submit Discharge Monitoring Reports to a state agency or EPA.

The Act requires use of best management practices for a wide range of other water discharges including nonpoint source pollution. Amendments in 1977 required stricter regulation of toxic pollutants. In 1987 Congress added NPDES permit coverage for municipal and industrial stormwater discharges.

## Land Pollution

Congress passed the Resource Conservation and Recovery Act (RCRA) in 1976, which created a regulatory framework for both municipal solid waste and hazardous waste disposed on land. RCRA requires that all hazardous wastes be managed and tracked from generation of the waste, through transport and processing, to final disposal, by means of a nationwide permit system. The *Hazardous and Solid Waste Amendments of 1984* mandated regulation of underground storage tanks containing petroleum and hazardous chemicals, and the phasing out of land disposal of hazardous waste. The Federal Facilities Compliance Act, passed in 1992, clarified RCRA coverage of federally owned properties such as military bases. Illegal disposal of waste is punishable by fines of up to $25,000 per occurrence.

## Noise Pollution

Passage of the Noise Control Act in 1972 established mechanisms of setting emission standards for virtually every source of noise including motor vehicles, aircraft, certain types of HVAC equipment and major appliances. It also put local government on notice as to their responsibilities in land use planning to address noise mitigation. This noise regulation framework comprised a broad data base detailing the extent of noise health effects. Congress ended funding of the federal noise control program in 1981, which curtailed development of further national regulations.

## State Programs

The state of California Office of Environmental Health Hazard Assessment (OEHHA) has maintained an independent list of substances with product labeling requirements as part of Proposition 65 since 1986.

## References


- Baechler, Christian; DeVuono, Matthew; Pearce, Joshua M. (2013). "Distributed Recycling of Waste Polymer into RepRap Feedstock". Rapid Prototyping Journal. 19 (2): 118–125. doi:10.1108/13552541311302978
- Dadd-Redalia, Debra (January 1, 1994). Sustaining the earth: choosing consumer products that are safe for you, your family, and the earth. New York: Hearst Books. p. 103. ISBN 978-0-688-12335-2
- "Report: "On the Making of Silk Purses from Sows' Ears," 1921: Exhibits: Institute Archives & Special Collections: MIT". mit.edu. Retrieved July 7, 2016
- Huesemann, M. H. (2003). "The limits of technological solutions to sustainable development" (PDF). Clean Techn Environ Policy. 5: 21–34. doi:10.1007/s10098-002-0173-8 (inactive 2017-01-24). Archived from the original (PDF) on 28 September 2011
- M.A. Kreiger, M.L. Mulder, A.G. Glover, J. M. Pearce, Life Cycle Analysis of Distributed Recycling of Post-consumer High Density Polyethylene for 3-D Printing Filament, Journal of Cleaner Production, 70, pp. 90–96 (2014). DOI: 10.1016/j.jclepro.2014.02.009
- Carl A. Zimring (2005). Cash for Your Trash: Scrap Recycling in America. New Brunswick, NJ: Rutgers University Press. ISBN 0-8135-4694-X
- Goodman, Peter S. (January 11, 2012). "Where Gadgets Go To Die: E-Waste Recycler Opens New Plant In Las Vegas". The Huffington Post. Retrieved July 29, 2015
- Brown, M. T.; Buranakarn, V. (2003). "Emergy indices and ratios for sustainable material cycles and recycle options" (PDF). Resources, Conservation and Recycling. 38 (1): 1–22. doi:10.1016/S0921-3449(02)00093-9
- Gunter, Matthew (January 1, 2007). "Do Economists Reach a Conclusion on Household and Municipal Recycling?". Econ Journal Watch. 4 (1): 83–111. Archived from the original on July 31, 2015
- Lynn R. Kahle; Eda Gurel-Atay, eds. (2014). Communicating Sustainability for the Green Economy. New York: M.E. Sharpe. ISBN 978-0-7656-3680-5
- Huesemann, M. H. (2003). "Recognizing the limits of environmental science and technology" (PDF). Environ. Sci. Technol. 37 (13): 259–261. Bibcode:2003EnST...37..259H. doi:10.1021/es0324930
- Morris, J. (2005). Comparative LCAs for curbside recycling versus either landfilling or incineration with energy recovery (12 pp). The International Journal of Life Cycle Assessment, 10(4), 273-284
- Unless otherwise indicated, this data is taken from The League of Women Voters (1993). The Garbage Primer. New York: Lyons & Burford. pp. 35–72. ISBN 1-55821-250-7
- Zaman, A. U.; Lehmann, S. (2011). "Challenges and opportunities in transforming a city into a "Zero Waste City"". Challenges. 2 (4): 73–93. doi:10.3390/challe2040073
- Clark, B.; Foster, J. B. (2009). "Ecological imperialism and the global metabolic rift : Unequal exchange and the guano/nitrates trade" (PDF). International Journal of Comparative Sociology. 50 (3–4): 311–334. doi:10.1177/0020715209105144. Archived from the original (PDF) on 27 April 2012
- Huesemann, M.; Huesemann, J. (2011). Techno-fix: Why Technology Won't Save Us or the Environment. New Society Publishers. p. 464. ISBN 978-0-86571-704-6. Retrieved 2016-07-07
- "A Beverage Container Deposit Law for Hawaii". www.opala.org. City & County of Honolulu, Department of Environmental Services. Oct 2002. Retrieved July 31, 2015

# PERMISSIONS

All chapters in this book are published with permission under the Creative Commons Attribution Share Alike License or equivalent. Every chapter published in this book has been scrutinized by our experts. Their significance has been extensively debated. The topics covered herein carry significant information for a comprehensive understanding. They may even be implemented as practical applications or may be referred to as a beginning point for further studies.

We would like to thank the editorial team for lending their expertise to make the book truly unique. They have played a crucial role in the development of this book. Without their invaluable contributions this book wouldn't have been possible. They have made vital efforts to compile up to date information on the varied aspects of this subject to make this book a valuable addition to the collection of many professionals and students.

This book was conceptualized with the vision of imparting up-to-date and integrated information in this field. To ensure the same, a matchless editorial board was set up. Every individual on the board went through rigorous rounds of assessment to prove their worth. After which they invested a large part of their time researching and compiling the most relevant data for our readers.

The editorial board has been involved in producing this book since its inception. They have spent rigorous hours researching and exploring the diverse topics which have resulted in the successful publishing of this book. They have passed on their knowledge of decades through this book. To expedite this challenging task, the publisher supported the team at every step. A small team of assistant editors was also appointed to further simplify the editing procedure and attain best results for the readers.

Apart from the editorial board, the designing team has also invested a significant amount of their time in understanding the subject and creating the most relevant covers. They scrutinized every image to scout for the most suitable representation of the subject and create an appropriate cover for the book.

The publishing team has been an ardent support to the editorial, designing and production team. Their endless efforts to recruit the best for this project, has resulted in the accomplishment of this book. They are a veteran in the field of academics and their pool of knowledge is as vast as their experience in printing. Their expertise and guidance has proved useful at every step. Their uncompromising quality standards have made this book an exceptional effort. Their encouragement from time to time has been an inspiration for everyone.

The publisher and the editorial board hope that this book will prove to be a valuable piece of knowledge for students, practitioners and scholars across the globe.

# Index

## O



## P



## R



## S



## T



## U



## V



## W



## Z